Polarized Sources and Targets

Tomohiro Uesaka
Hideyuki Sakai
University of Tokyo, Japan

Akihiro Yoshimi
Koichiro Asahi
RIKEN, Japan

PROCEEDINGS OF THE
11TH INTERNATIONAL WORKSHOP ON
Polarized Sources
and Targets

Tokyo, Japan   14 – 17 November 2005

World Scientific

NEW JERSEY • LONDON • SINGAPORE • BEIJING • SHANGHAI • HONG KONG • TAIPEI • CHENNAI

Published by
World Scientific Publishing Co. Pte. Ltd.
5 Toh Tuck Link, Singapore 596224
USA office: 27 Warren Street, Suite 401-402, Hackensack, NJ 07601
UK office: 57 Shelton Street, Covent Garden, London WC2H 9HE

British Library Cataloguing-in-Publication Data
A catalogue record for this book is available from the British Library.

POLARIZED SOURCES AND TARGETS
Proceedings of the 11th International Workshop

Copyright © 2007 by World Scientific Publishing Co. Pte. Ltd.

All rights reserved. This book, or parts thereof, may not be reproduced in any form or by any means, electronic or mechanical, including photocopying, recording or any information storage and retrieval system now known or to be invented, without written permission from the Publisher.

For photocopying of material in this volume, please pay a copying fee through the Copyright Clearance Center, Inc., 222 Rosewood Drive, Danvers, MA 01923, USA. In this case permission to photocopy is not required from the publisher.

ISBN-13 978-981-270-703-1
ISBN-10 981-270-703-4

Printed in Singapore.

Preface

The 11th Workshop on Polarized Sources and Targets was held on November 14-17, 2005 at the Ichijo hall, Yayoi Auditorium of the University of Tokyo, Japan. The workshop was co-hosted by RIKEN and the Center for Nuclear Study, the University of Tokyo. It was supported by Research Center for Nuclear Research, Osaka University and the international spin physics committee.

The workshop is a traditional one to discuss physics and technologies related to the polarized gas/solid targets, polarized electron/ion/neutron sources, and polarimetry.

80 scientists from different institutes attended the workshop and discussed about state-of-the-art techniques in spin polarization. It provided a good opportunity for scientists in the related fields to exchange ideas and information on the recent progress of adjacent fields. For the above purpose, the workshop program was organized from the view point of the technical methods, but not of applications, as listed below.

- Atomic Beam Method
- Optical Pumping Method and Laser Techniques
- NMR/ESR Method
- Nuclear Reaction Method
- Cryogenic Technique
- Other Techniques

In addition to sessions on these long-standing topics of the field, special sessions on utilization of the polarization technique in studies of unstable nuclei were arranged. Two stimulating proposals on new polarization schemes were also presented in the workshop.

The organizing committee would like to thank the speakers for their exciting talks and the participants for their active discussions. We appreciate our staffs and graduate students of the University of Tokyo and RIKEN for their invaluable help in preparing the workshop.

H. Sakai, K. Asahi, T. Uesaka, and A. Yoshimi

Local Organizing Committee

K. Asahi (co-chair, RIKEN/TITech), H. Sakai (co-chair, CNS Tokyo),
H. En'yo (RIKEN), K. Hatanaka (RCNP), N. Horikawa (Chubu),
K. Imai (Kyoto), T. Iwata (Yamagata), T. Kawabata (CNS Tokyo),
Y. Miyachi (TITech), Y. Mori (KEK), T. Nakanishi (Nagoya),
H. Okamura (Tohoku), N. Sakamoto (RIKEN), Y. Sakemi (RCNP),
T. Shibata (TITech), T. Shimoda (Osaka), T. Tamae (Tohoku),
A. Tamii (RCNP), M. Uchida (TITech), T. Uesaka (scientific secretary,
CNS Tokyo), K. Yako (Tokyo), A. Yoshimi (scientific secretary, RIKEN)

International Advisory Committee

T. Roser (chair, BNL), A.D. Krisch (pst-chair, Michigan),
F. Bradamante (Trieste), O. Chamberlain* (Berkeley),
E.D. Courant* (BNL), D.G. Crabb (Virginia), A.V. Efremov (JINR),
G. Fidecaro* (CERN), W. Haeberli* (Wisconsin), K. Hatanaka (RCNP),
K. Imai (chair-Elect, Kyoto), G. Mallot (CERN), R.G. Milner (MIT),
Y. Mori (KEK), C.Y. Prescott (SLAC), F. Rathmann (COSY),
Y.M. Shatunov (Novosibirsk), V. Soergel* (Heidelberg),
L.D. Soloviev* (IHEP), E. Steffens (Erlangen), E.J. Stephenson (Indiana),
N.E. Tyurin (IHEP), W.T.H. van Oers* (Manitoba)

* Honorary Members

PST05 Program

11/14 (Mon)

Morning Session
10:30	H. Sakai (Tokyo)	Welcome
10:35	K. Asahi (RIKEN)	Opening Address
10:40	A. Masaike (JSPS)	Historical Review on Polarized Targets
11:20	T. Wise (Madison)	The RHIC Polarized Hydrogen Jet Target: experience and future prospects

Lunch

Afternoon Session I — Cryogenic Method I —
13:40	W. Meyer (Bochum)	SPIN04 Summary & Progresses in DNP Targets
14:25	A. Beda (ITEP)	The aligned nuclear targets from Sb, In and I_2 for investigation of Time Reversal Invariance Violation
14:45	N. Doshita (Bochum)	Future activities of the COMPASS polarized target

15:05 *Coffee Break*

Afternoon Session II — Nuclear Polarization in RIB experiments I —
15:35	T. Wakui (Tohoku)	Polarized Proton Solid Target for RI beam experiments
16:05	T. Furukawa (Osaka)	Laser-microwave double resonance method in superfluid helium for measurement of nuclear moments
16:25	T. Shimoda (Osaka)	Polarized ^{11}Li beam at TRIUMF and its application for spectroscopic study of the daughter nucleus ^{11}Be
16:45	P. Hautle (PSI)	Polarized solid targets at PSI: recent developments

17:10 *Poster Session & Reception*

11/15 (Tue)

Morning Session I — Nuclear Reaction Method—

9:00	N. Saito (Kyoto)	Production of Polarized Neutron at RHIC (tentative)
9:30	K. Yako (Tokyo)	Polarimeters for a test of EPR paradox
9:50	V. P. Ladygin (JINR)	Deuteron beam polarimetry at Nuclotron
10:10	E. J. Stephenson (IUCF)	Deuteron Polarimeter for Electric Dipole Moment Search
10:30	*Coffee Break*	

Morning Session II — Polarized Electron Beam I —

11:00	M. Poelker (JLab)	Operation of CEBAF Photoguns at Average Beam Current > 1 mA
11:30	J.E. Clendenin (SLAC)	ILC SLAC R&D Program for a Polarized RF Gun
11:50	M. Yamamoto (Nagoya)	High field gradient polarized electron gun for ILC

Lunch

Afternoon Session I — Cryogenic Method II—

13:30	D.G.Crabb (Virginia)	Proton and Deuteron Polarizations with Irradiated Materials
13:55	C. Djalai (South Carolina)	Magnet and beam-target interaction studies for the Jlab Hall-B frozen spin polarized target
14:15	Y. Kisselev (CERN)	Microwave Cavity for large COMPASS polarized target
14:35	S. Bouchigny (Orsay)	Distillation and Polarization of HD
14:55	T. Kageya (BNL)	Performances of frozen-spin polarized HD targets for nucleon spin experiments
15:15	*Coffee Break*	

Afternoon Session II — Optical Pumping Method I —

15:45	M. Romalis (Princeton)	Tests of Lorentz symmetry and other applications of noble-gas alkali-metal co-magnetometers
16:15	Y. Masuda (KEK)	A new ^{3}He polarization for fundamental neutron physics
16:35	B. Clasie (MIT)	The MIT laser driven target of high-density, nuclear polarized hydrogen gas
16:55	V. Fimushkin (JINR)	A proposal of a polarized ^{3}He^{++} ion source with penning ionizer for JINR

11/16 (Wed)

Morning Session I — Atomic Beam Method—

9:00	E.P. Tsentalovich (MIT)	Polarized internal gas target in a strong toroidal magnetic field
9:30	B. Juhasz (Stefan Meyer)	An atomic beam line to measure the ground-state hyperfine splitting of antihydrogen
9:50	R. Engels (FZ Juelich)	The polarized internal gas target of ANKE at COSY

15:15 *Coffee Break*

Morning Session II — Nuclear Polarization in RIB experiments II—

10:40	H. Ueno (RIKEN)	Production of spin-oriented unstable nuclei via the projectile-fragmentation reaction
11:10	G. Goldring (Weizmann)	Tilted Foil Nuclear Polarization
11:30	M. Mihara (Osaka)	Spin polarization of ^{23}Ne produced in heavy ion reactions

Workshop Photo, Lunch, Excursion and Banquett

11/17 (Thu)

Morning Session I — Optical Pumping Method II—

9:00	A. Zelenski (BNL)	Polarized proton beams in RHIC
9:25	I. Krimmer (Mainz)	Polarized ^{3}He targets at MAMI-C
9:45	Y. Shimizu (RCNP)	Development of the Polarized ^{3}He Target at RCNP
10:15	A. Tamii (RCNP)	Design of a polarized ^{6}Li^{3+} ion source and its feasibility

10:35 *Coffee Break*

Morning Session II — Polarized Electron Beam II—

11:05	M. Farkhondeh (MIT-Bates)	Polarized electron sources for future electron-hadron colliders
11:25	K. Ioakeimidi (SLAC)	Comparison of AlInGaAs/GaAs superlattice photocathodes having low conduction band offset
11:45	M. Kuwahara (Nagoya)	Generation of polarized electrons by field emission

Lunch

Afternoon Session I — New Methods —

13:30	T. Nakajima (Kyoto)	Spin-polarization using optical methods
14:00	T. Iwata (Yamagata)	An attempt toward dynamic nuclear polarization for liquid He^3

Closing

Poster Presentations

A. Brachmann (SLAC) — Performance of GaAsP/GaAs Superlattice Photocathodes in High Energy Experiments Using Polarized Electrons

K. Itoh (Saitama) — Development of Spin-Exchange Type Polarized ^{3}He Target for RI-Beam Experiments

D. Kameda (TITech) — Production of spin-polarized RI beams via projectile fragmentation reaction and the application to nuclear moment measurements

T. Katabuchi (Gunma) — A new ^{3}He polarizer and target system for low-energy scattering Measurements

J. Koivuniemi (CERN) — Polarization data analysis of the COMPASS ^{6}LiD target

T. Nakanishi(Nagoya) — Review of semiconductor photocathodes developed for highly polarized electron source

I. Nishikawa (Tohoku) — Lamb-shift polarimeter for deuteron gas target at LNS

S. Noji & K. Miki (Tokyo) — Performance Evaluation of Neutron Polarimeter NPOL

M. Poelker (JLab) — Status of Polarized Beam Operations at Jefferson Lab

A. Raccanelli (Bonn) — Current developments of the Bonn polarized solid target

S. Sakaguchi (Tokyo) — Polarization Measurement of Polarized Proton Solid Target via $\vec{p}+^4$He Elastic Scattering

K. Suda (Tokyo) — A new tool to calibrate deuteron beam polarization at intermediate energies

M. Tanaka (Tokiwa) — Polarized ^{3}He ion source based on the spin-exchange collisions

M. Yamaguchi (RIKEN) — Extraction of fractions of the resonant component from analyzing powers in ^{6}Li$(d,\alpha)^4$He and ^{6}Li$(d,p_0)^7$Li reaction at very low incident energies

Contents

Review talk

Historical Review on Polarized Target 1
A. Masaike

Atomic Beam Method

RHIC polarized hydrogen jet target:
experience and future prospects 11
T. Wise

Polarized internal gas target in a strong toroidal magnetic field 18
E. Tsentalovich, E. Ihloff, H. Kolster, N. Meitanis R. Milner,
A. Shinozaki, V. Ziskin, Y. Xiao, C. Zhang

An atomic beam line to measure the ground-state hyperfine
splitting of antihydrogen 24
B. Juász, E. Widmann, D. Barna, J. Eades, R.S. Hayano, M. Hori,
W. Pirkl, D. Horváth, T. Yamazaki

Installation and commissioning of the polarized internal gas
target of the magnet spectrometer ANKE at COSY-Jülich 28
R. Engels, D. Chiladze, S. Dymov, K. Grigoryev, D. Gusev,
B. Lorentz, D. Prasuhn, F. Rathmann, J. Sarkadi, H. Seyfarth,
H. Stroher

Cryogenic Method

Progress in dynamically polarized solid targets 35
J. Heckmann, S. Goertz, Chr. Hess, W. Meyer, E. Radtke,
G. Reicherz

The Sb, LiIO$_3$ and HIO$_3$ aligned nuclear targets for investigation
of time reversal invariance violation 45
A.G. Beda, L.D. Ivanova

Future activities of the COMPASS polarized target 50
N. Doshita, J. Heckmann, Ch. Hess, Y. Kisselev, J. Koivuniemi,
K. Kondo, W. Meyer, G. Reicherz

Polarizations in irradiated proton and deuterated materials 54
D.G. Crabb

Magnet and beam-target interaction studies for the Jlab Hall-B
frozen spin polarized target 59
O. Dzyubak, C. Djalali, S. Strauch, D. Tedeschi

Investigations of the multimode cavity for the COMPASS-magnet 63
Y. Kisselev, N. Doshita, J. Heckmann, J. Koivuniemi, K. Kondo,
W. Meyer, G. Reicherz, G. Baum, F. Gautheron, J. Ball,
A. Magnon, S. Platchkov, G. Mallot

Distillation and polarization of HD 67
S. Bouchigny, J-P. Didelez, G. Rouille

Performance of frozen-spin polarized HD targets for nucleon
spin experiments 72
T. Kageya, K. Ardashev, C. Bade, M. Blecher, A. Caracappa,
A. D'Angelo, A. D'Angelo, R. Di Salvo, A. Fantini, C. Gibson,
H. Glückler, K. Hicks, S. Hoblit, A. Honig, M. Khandakar,
S. Kizigul, O. Kistner, S. Kucuker, A. Lehmann, F. Lincoln,
R. Lindgren, M. Lowry, M. Lucas, J. Mahon, L. Miceli,
D. Moricciani, B. Norum, M. Pap, B. Preedom, A.M. Sandorfi,
C. Schaerf, H. Seyfarth, H. Ströher, C. Thorn, K. Wang, X. Wei,
C. Whisnant

Optical Pumping Method

A new ^{3}He polarization for fundamental neutron physics 79
Y. Masuda, T. Ino, S.C. Jeong, S. Muto, Y. Watanabe, V.R. Skoy

The MIT laser-driven target of nuclear polarized hydrogen gas 84
B. Clasie, C. Crawford, D. Dutta, H. Gao, J. Seely, W. Xu

A proposal of polarized ^{3}He^{++} ion source with Penning ionizer
for JINR 88
*N.N. Agapov, N.A. Bazhanov, Yu.N. Filatov, V.V. Fimushkin,
L.V. Kutuzova, V.A. Mikhailov, Yu.A. Plis, Yu.V. Prokofichev,
V.P. Vadeev*

Polarized proton beams in RHIC 93
A. Zelenski, for the RHIC Spin Collaboration

Polarized ^{3}He targets at MAMI-C 99
J. Krimmer

Development of a polarized ^{3}He target at RCNP 103
*Y. Shimizu, K. Hatanaka, A.P. Kobushkin, T. Adachi, K. Fujita,
K. Itoh, T. Kawabata, T. Kudoh, H. Matsubara, H. Ohira,
H. Okamura, K. Sagara, Y. Sakemi, Y. Sasamoto, Y. Shimbara,
H.P. Yoshida, K. Suda, Y. Tameshige, A. Tamii, M. Tomiyama,
M. Uchida, T. Uesaka, T. Wakasa, T. Wakui*

Design of a polarized ^{6}Li^{3+} ion source and simulations of feasibility .. 107
*A. Tamii, K. Hatanaka, H. Okamura, Y. Sakemi, Y. Shimizu,
K. Fujita, Y. Tameshige, H. Matsubara*

Nuclear Reaction Method

Focal plane polarimeter for a test of EPR paradox 113
*K. Yako, T. Saito, H. Sakai, H. Kuboki, M. Sasano,
T. Kawabata, Y. Maeda, K. Suda, T. Uesaka, T. Ikeda, K. Itoh,
N. Matsui, Y. Satou, K. Sekiguchi, H. Matsubara, A. Tamii*

Deuteron beam polarimetry at nuclotron 117
*V.P. Ladygin, L.S. Azhgirey, Yu.V. Gurchin, A.YU. Isupov,
M. Janek, J.-T. Karachuk, A.N. Khrenov, A.S. Kiselev,
V.A. Kizka, V.A. Krasnov, A.N. Livanov, A.I. Malakhov,
V.F. Peresedov, Yu.K. Pilipenko, S.G. Reznikov, T.A. Vasiliev,
V.N. Zhmyrov, L.S. Zolin, T. Uesaka, T. Kawabata, Y. Maeda,
S. Sakaguchi, H. Sakai, Y. Sasamoto, K. Suda, K. Itoh,
K. Sekiguchi, I. Turzo*

Deuteron polarimeter for electric dipole moment search 121
E.J. Stephenson

Polarized Electron Production

Operation of CEBAF photoguns at average beam current >1 mA 127
M. Poelker, J. Grames

ILC SLAC R&D program for a polarized RF gun 134
*J.E. Clendenin, A. Brachmann, D.H. Dowell, E.L. Garwin,
K. Ioakeimidi, R.E. Kirby, T. Maruyama, R.A. Miller,
C.Y. Prescott, J.W. Wang, J.W. Lewellen, R. Prepost*

High field gradient polarized electron gun for ILC 138
*M. Yamamoto, N. Yamamoto, T. Nakanishi, S. Okumi,
M. Kuwahara, K. Yasui, T. Morino, R. Sakai, K. Tamagaki,
F. Furuta, M. Kuriki, H. Matsumoto, M. Yoshioka*

Polarized electron sources for future electron ion colliders 142
*M. Farkhondeh, W. Franklin, E. Tsentalovich, Ilan Bem-Zvi,
V. Litvinenko*

Comparison of AlInGaAs/GaAs superlattice photocathodes having
low conduction band offset 147
*K. Ioakeimidi, T. Maruyama, J.E. Clendenin, A. Brachmann,
E.L. Garwin, R.E. Kirby, C.Y. Prescott, D. Vasilyev,
Y.A. Mamaev, L.G. Gerchikov, A.V. Subashiev, Y.P. Yashin*

Generation of spin polarized electrons by field emission 152
*M. Kuwahara, T. Nakanishi, S. Okumi, M. Yamamoto,
M. Miyamoto, N. Yamamoto, K. Yasui, T. Morino, R. Sakai,
K. Tamagaki, K. Yamaguchi*

Polarization in RI beam experiments

Polarized proton solid target for RI beam experiments 159
T. Wakui

Laser-microwave double resonance spectroscopy in superfluid
helium for the measurement of nuclear moments 165
*T. Furukawa, T. Shimoda, Y. Matsuo, Y. Fukuyama, T. Kobayashi,
A. Hatakeyama, T. Itou, Y. Ota*

Polarized ^{11}Li beam at TRIUMF and its application for
spectroscopic study of the daughter nucleus ^{11}Be 169
*T. Shimoda, Y. Hirayama, H. Izumi, Y. Akasaka, K. Kawai,
I. Wakabayashi, M. Yagi, Y. Yano, A. Hatakeyama, C.D.P. Levy,
K.P. Jackson, H. Miyatake*

Polarised solid targets at PSI: recent developments 173
*B. van den Brandt, P. Hautle, J.A. Konter, F.M. Piegsa,
J.P. Urrego-Blanco*

Production of spin-oriented unstable nuclei via the projectile-
fragmentation reaction 178
*H. Ueno, D. Kameda, A. Yoshimi, T. Haseyama, K. Asahi,
M. Takemura, G. Kijima, K. Shimada, D. Nagae, M. Uchida,
T. Arai, S. Suda, K. Takase, T. Inoue*

Titled foil nuclear polarization 184
G. Goldring

Spin polarization of ^{23}Ne produced in heavy ion reactions 188
*M. Mihara, R. Matsumiya, K. Matsuta, T. Nagatomo, M. Fukuda,
T. Minamisono, S. Momota, Y. Nojiri, T. Ohtsubo, T. Izumikawa,
A. Kitagawa, M. Torikoshi, M. Kanazawa, S. Sato, J.R. Alonso,
G.F. Krebs, T.J.M. Symons*

New Methods

Spin-polarization using optical methods 195
T. Nakajima

An attempt toward dynamic nuclear polarisation for liquid ^{3}He 201
*T. Iwata, S. Kato, H. Kato, T. Michigami, T. Nomura, T. Shishido,
Y. Tajima, H. Ueno, H.Y. Yoshida*

Poster presentations

Performance of GaAsP/GaAs superlattice photocathodes in
high energy experiments using polarized electrons 207
    *A. Brachmann, J.E. Clendenin, T. Maruyama, E.L. Garwin,
K. Ioakeimidi, C.Y. Prescott, J.L. Turner, R. Prepost*

Development of spin-exchange type polarized ^{3}He target for
RI-beam experiments 209
    K. Itoh

Production of spin-polarized RI beams via projectile fragmentation
and the application to nuclear moment measurements 211
    *D. Kameda, H. Ueno, K. Asahi, A. Yoshimi, H. Watanabe,
T. Haseyama, Y. Kobayashi, M. Uchida, H. Miyoshi, K. Shimada,
G. Kijima, M. Takemura, D. Nagae, G. Kato, S. Emori, S. Oshima,
T. Arai, M. Tsukui*

A new ^{3}He polarizer and target system for low-energy
scattering measurements 213
    T. Katabuchi, T. B. Clegg, T.V. Daniels, H.J. Karwowski

Polarization data analysis of the COMPASS ^{6}LiD target 215
    *J. Koivuniemi, N. Doshita, Y. Kisselev, K. Kondo, W. Meyer,
G. Reicherz, F. Gautheron*

Lamb-shift polarimeter for deuteron gas target at LNS 217
    I. Nishikawa, T. Tamae

Performance evaluation of NPOL at RIKEN 219
    *S. Noji, K. Miki K. Yako, H. Sakai, H. Kuboki, T. Kawabata,
K. Suda, K. Sekiguchi*

A frozen-spin target for the TOF detector 221
    A. Raccanelli, H. Dutz, R. Krause

Polarization measurement of polarized proton solid target via
the $\vec{p}+^4$He elastic scattering 223
S. Sakaguchi, T. Uesaka, T. Wakui, T. Kawabata, N. Aoi,
Y. Hashimoto, M. Ichikawa, Y. Ichikawa, K. Itoh, M. Itoh,
H. Iwasaki, T. Kawahara, H. Kuboki, Y. Maeda, R. Matsuo,
T. Nakao, H. Okamura, H. Sakai, N. Sakamoto, Y. Sasamoto,
M. Sasano, Y. Satou, K. Sekiguchi, M. Shinohara, K. Suda,
D. Suzuki, Y. Takahashi, A. Tamii, K. Yako, M. Yamaguchi

A new tool to calibrate deuteron beam polarization at
intermediate energies 225
K. Suda, H. Okamura, T. Uesaka, J. Nishikawa, H. Kumasaka,
R. Suzuki, H. Sakai, A. Tamii, T. Ohnishi, K. Sekiguchi, K. Yako,
S. Sakoda, H. Kato, M. Hatano, Y. Maeda, T. Saito, T. Ishida,
N. Sakamoto, Y. Satou, K. Hatanaka, T. Wakasa, J. Kamiya

Polarized ^{3}He ion source based on the spin-exchange collisions 227
M. Tanaka, Y. Takahashi, T. Shimoda, and T. Furukawa, S. Yasui,
M. Yosoi, K. Takahisa

Extraction of fractions of the resonant component from analyzing
powers in ^{6}Li(d, α)^{4}He and ^{6}Li(d, p$_0$)^{7}Li reaction at very low
incident energies 230
M. Yamaguchi, Y. Tagishi, Y. Aoki, T. Iizuka, T. Nagatomo,
T. Shinba, N. Yoshimaru, Y. Yamato, T. Katabuchi, M. Tanifuji

Participants list 233

Historical Review on Solid Polarized Targets

AKIRA MASAIKE
Washington Center, Japan Society for the Promotion of Science

Protons in the crystal of $La_2Mg_3(NO_3)_{12}24H_2O$ were found to be polarized dynamically by means of the solid effect at the beginning of 1960's. Organic materials were polarized up to 80 % in ^{3}He cryostats in 1969. It can be interpreted as the equal spin temperature model. The spin frozen targets with dilution refrigerators were constructed in 1974. Such targets have been used for experiments which require wide access angles. The NH_3, LiH and LiD were polarized in 1980's. Static polarization of HD has been tried and proton polarization higher than 60 % was obtained with small amount of ortho H_2 in 2000. We mention also the dynamic polarization of aromatic molecules in high temperature.

1. Introduction - Orientation of Nuclear Spin

An assembly of nuclei with magnetic moments in matter in thermal equilibrium at the temperature T and magnetic field B orient themselves in the direction of the field. The degree of nuclear polarization P_n is the average $<I_z>/I_z$ taken over all the spins of the nuclei in the sample, where I is the nuclear spin and I_z is the magnetic sub-state of the nuclear spin. For a spin $I=1/2$, P_n is given by

$$P_n = \tanh(\mu B / kT). \tag{1-1}$$

The B and μ are the external magnetic field and the nuclear magnetic moment, respectively. It is

$$P_n = \tanh(1.02 \times 10^{-7} B/T) \tag{1-2}$$

for protons.

Whereas, the degree of nuclear alignment A_n is defined as

$$A_n = <I_z^2 - I(I+1)/3>/I^2 \tag{1-3}$$

The nuclear alignment of proton is zero, since $I=1/2$.

It is desirable to get high polarization of nucleons for high energy spin physics. For this purpose, low temperature and high magnetic field are indispensable. The proton polarization of 76% could be obtained at 0.01K and in 10T, if the polarization buildup time is sufficiently short. Such a method for polarizing the nucleons statically is called as " the brute force method".

On the other hand, we can get sizable polarizations of nuclei in paramagnetic materials by dynamical methods in rather lower field and at higher temperature than in the case of the brute force method. In the dynamical method the magnetic coupling between electron spins and nuclear spins are used to transfer the polarization of electrons to nuclei. The method was proposed originally by Overhauser, who predicted that the saturation of spin resonance of conduction electrons in metals could lead to a nuclear polarization comparable to the electronic polarization [1]. Then, it was shown by Abragam that the similar method could be extended to non-metalic substances, in particular solids containing paramagnetic impurities [2]. He showed also that the paramagnetic center and the nucleus being polarized need not to belong to the same atom. This phenomenon was named as " the solid effect".

2. LMN Target

The energy level diagram of a paramagnetic center coupled to a single neighboring proton in high external magnetic field is shown in Fig.1(a).

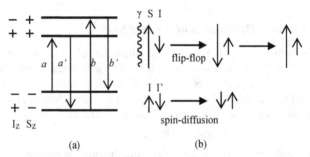

Figure 1 (a). Energy level diagram of the paramagnetic center coupled to the proton.
(b). Flip-flop of an electron spin and a proton spin. The nuclear spin diffusion is also shown.

The splitting by the electron Zeeman interaction is larger than that by the proton Zeemen interaction by factor 600. Transitions a and b in Fig. 1(a) are the "forbidden transitions" in which both S_z and I_z reverse the sign. On the other hand, transitions a' and b' in which S_z reverses and I_z remains the same are called as "the allowed transition". The transition probabilities for the forbidden transitions are much smaller than these for the allowed transitions. However, the dipole-dipole coupling does ensure that there is a small admixture of $S_z= +1/2$ and -1/2. The admixture makes forbidden transitions possible by the RF field which is applied at the frequency corresponding to the energy difference between two levels. The populations of the two levels connected by the forbidden transition can be equalized. It corresponds to the "flip-flop" of an electron spin and a proton spin as shown in Fig. 1 (b). Because of the strong coupling between electrons and lattice, the relaxation time of electron (T_{1s}) is short. It corresponds to the allowed transition. Whereas, the proton spin stays on higher level for long time because of the weakness of coupling between the proton spin and the lattice. Such processes clearly provide a means for the orientation of the proton spin. As two neighboring protons are coupled by the dipole-dipole interaction, the orientation of proton spin diffuses throughout whole material, trending to equalize the nuclear polarization.

At the beginning of the history of dynamically polarized target, protons in the crystal of $La_2Mg_3(NO_3)_{12}24H_2O$ (LMN), containing a small percent (e.g. 0.2%) of neodymium, was polarized by Abragam and Jeffries [3], [4].

Since the NMR signals for positive and negative polarizations are well-resolved for LMN crystal as shown in Fig. 2, the solid effect works so well. The proton polarization of about 80 % was obtained in the temperature around 1 K and the magnetic field of 1.8 T. LMN targets were successfully operated for elastic scattering experiments with π, K, p and n beams.

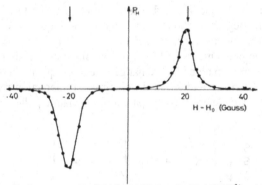

Figure 2. NMR amplitude vs. magnetic field for LMN doped with 1 % Nd^{3+} at 4.2 K and in 1.95 T

However, the target with LMN crystal is not convenient for other reactions than elastic scattering because of the enormous amount of background events related to nuclei different from hydrogen, since the dilution factor (polarizable nucleons/total number of nucleons) is too small. For example, the measurement of the asymmetry in $\pi^+ + p \rightarrow K^+ + \Sigma^+$ reaction was carried out using LMN targets at CERN and Berkeley in 1964 for confirmation of parity conservation in strong interaction, but it was not so successful.

In addition, as the crystal is damaged seriously with relativistic particles of $2 \times 10^{12}/cm^2$, it is difficult to use the target for electron and photon beams.

3. Organic Materials in ^{3}He Cryostat

Organic materials with free radicals had been tried to polarize dynamically in many laboratories in the last half of 1960's, since these materials have higher concentrations of free protons and are stronger for radiation damage than LMN. More than 100 materials had been tested with several kinds of free radicals at

CERN [5]. Despite the tremendous efforts had been made, none of materials had been successfully polarized up to 30 % until 1968.

Major breakthroughs were achieved dramatically in 1969. Protons in butanol with small amount of water doped with porphyrexide were polarized up to 40 % at 1 K and in 2.5 T by Mango et al. at CERN [6]. In the meantime, protons in diol with Cr^{5+} complex were polarized up to 45 % at 1 K and in 2.5 T by Glättli et al. at Saclay [7]. A few months later protons in diol was found to be polarized up to 80 % at the temperature lower than 0.5 K by Masaike et al. at Saclay [8], while Argonne group (Hill et al.) succeeded in polarizing protons in butanol up to 67 % in a ^{3}He cryostat [9]. These values are surprisingly large, since it had been believed that the polarization at lower temperature than 1 K would be less than that at 1 K before their trials.

Electron spin resonance lines are so broad in organic materials compared to the nuclear Larmor frequency, and it is difficult to separate the contributions from positive and negative solid effects. Microwave frequencies for positive and negative polarizations are far more separated (~ 0.01T for peak to peak) than expected, and the phenomena could not be explained by the simple solid effect. The phenomena are explained by exchange of energy quanta between a nuclear Zeeman energy reservoir and an electron spin-spin interaction reservoir. In this model more than two electrons participate to the dynamic polarization.

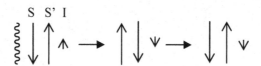

Figure 3. Cooling of the nuclear spin by the electron spin-spin interaction reservoir.

It was found that deuterons in deuterated organic materials can also be polarized in the same way. The relation between the polarizations of protons and deuterons can be interpreted as " the equal spin temperature model" by writing $P = B_I (\mu B/2kT_s)$, where B_I is the Brillouin function and T_s is the nuclear spin temperature.

Diols and butanol had been found to be damaged with relativistic particles of ~ 5×10^{14}/cm^2, which is 250 times as much as the ones in the case of LMN. Therefore, these targets could be used for electron and photon beams.

From 1970, polarized targets with diols and butanol cooled in ^{3}He cryostats have been used in most of the laboratories and the LMN target disappeared.

4. Spin Frozen Target

In 1965 Schmugge and Jeffries discussed the possibility of maintaining the polarization without microwave irradiation, if the nuclear spin relaxation time is long enough [10]. Such a target is advantageous because of the large access angle around the target area in less homogeneous and lower magnetic field than in the polarizing condition. It was constructed in Rutherford laboratory for the first time and named as the "spin frozen target" by Russell [11]. After polarizing at 0.5 K and in 5 T, the target was held at 0.3 K and in 3 T. The relaxation time of protons in glycerol-water mixture was about 33 hrs.

The relaxation time of proton is more than 2 weeks at 100 mK and in 1 T. Such targets were realized through the development of dilution refrigerators at CERN and KEK in 1974. The dilution refrigerator at CERN had continuous flow heat exchangers with sintered cooper [12]. The maximum beam intensity which was limited by Kapitza resistance was of the order of 10^8 particles/sec.

A spin frozen deuteron target was constructed at KEK for the reactions $K^+n \rightarrow K^+n$, K^0p almost at the same time [13]. Target material was deuterated propanediol doped with Cr^{5+} complex. A typical polarization was 40 % with frequency modulation. The relaxation time was about 2 weeks at 80 mK.

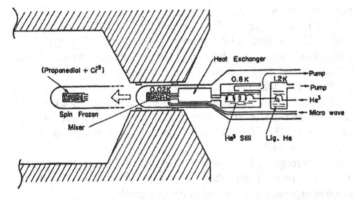

Figure 4. Schematic view of the spin frozen deuteron target at KEK

The spectrometer consists of a C-type magnet with a large gap and a pair of small rectangular pole pieces attached on the pole surface of the magnet. The target was inserted in the small pole gap situated in the spectrometer. After polarized, the target was moved downstream the beam from the center of the small pole gap and was used for the scattering experiment at this position.

Saclay group constructed a spin frozen target in late 1970's for nucleon-nucleon scattering experiments [14]. It consisted of a high power dilution refrigerator for target material of 100 cm^3, a vertical polarizing magnet and two holding magnets. Spin frozen targets were also constructed in Bonn, PSI, Dubna and other laboratories.

5. NH_3, LiH and LiD Targets

Ammonia is advantageous as target material because of the high dilution factor (0.176). In the early trial of dynamic polarization, the proton polarization of

70 % was obtained in NH_3 doped with ethylene glycol Cr^{5+} complex at 0.5 K and in 2.5 T by Scheffler and Borghini at CERN [15]. However, it was not easy to handle ammonia. In particular, difficulties were in reproducibility of DNP and slow growth of the polarization. Whereas, the polarization of irradiated ammonia was reported in CERN in 1979. They irradiated ammonia with 0.95×10^{15} protons/cm^2 from a 580 MeV proton beam line in liquid N_2 and obtained the proton polarization of 90~93 % with radicals of about 5×10^{18} spins/cm^3. However, they found that further proton irradiation led to explosions [16]. Shortly after the trial at CERN, it was found by Meyer et al. at Bonn that ammonia could be prepared by irradiation of 10^{17} e/cm^3 in liquid argon at about 90 K without explosion. They got the polarization of ≥90 % at 0.3 K and in 2.5 T for protons and ~ 40 % for deuterons. The polarization buildup time is short enough and the material is much stronger for the radiation damage than diols and butanol. In addition, it was chemically stable for more than a year.

After the success in Bonn, ammonia irradiated in liquid argon became one of the popular polarized targets.

LiH and LiD are very useful materials for the polarized target, since they have higher dilution factors than ammonia. The dilution factor of ^{6}LiD is 0.5.

In the early stage of the history of dynamic polarization, Abragam group used ^{6}LiF to validate the Overhauser effect in metals in 1960 [17].

In 1969, ^{6}LiF was used for studying on anti-ferromagnetism in Saclay. Then, LiH crystal irradiated with electrons was proposed as a target material in 1978, and polarizations of 80 % for ^{7}Li and 35 % for ^{6}Li were obtained at 200 mK and in 6.5 T [18], [19]. The results are in good agreement with the equal spin temperature model.

Then, the groups in PSI and Bonn confirmed the high polarization of H, D and ^{6}Li in 2.5 T. In particular, the best conditions to polarize ^{6,7}LiH and ^{6}LiD in the dilution refrigerator were found in Bonn recently.

The COMPASS experiment is now going on at CERN to investigate the spin structure of nucleons [20]. The COMPASS target consists of ^{6}LiD of 350 grs. which is irradiated with 20 MeV electrons. The polarization of ^{6}Li and D are about 50 % at 300 mK and in 2.5 T. There are two targets polarized in the opposite directions in a same mixing chamber in order to cancel the false asymmetry. The holding temperature is ~ 60 mK and the nuclear relaxation time is 1,500 hrs. in 0.42 T.

6. Hydrogen Deuteride (HD)

Hydrogen deuteride is potentially a highly advantageous polarized target, since in principle all the nucleons in the crystal are polarizable.

It was difficult to obtain low temperature and high field for obtaining high polarization by the brute force method in 1960's. However, longtime efforts have paid off by many researchers and now it is within our reach.

Hardy and Honig proposed the HD polarization by the brute force method in the middle of 1960's [21]. They measured the relaxation properties of the proton and deuteron at 0.5 K, which depend on the ortho-para and para-ortho conversions of H_2 and D_2, respectively. Honig pointed out that if protons and deuterons are polarized with a little ortho H_2 at 0.5 K, the polarization can be kept for long time at the ^{4}He temperature after ortho H_2 converts to para H_2. It took about 2 months for the ortho-para conversion.

However, 35 years have passed without any actual polarized target of HD.

In recent years, Grenoble-Orsay group succeeded in polarization of protons and deuterons of $\geq$ 60 % and $\geq$ 14 %, respectively, by means of the brute force method at 10 mK and in 13.5 T. Small concentrations of ortho H_2 and para D_2 were necessary for polarization [22]. While, LEGS group in BNL obtained 70 % and 17 % polarizations for protons and deuterons, respectively, at 17 mK and in 15 T. They held the polarization at 1.25 K and in 0.7 T [23].

7. Polarization of Organic Materials with Pentacene Molecules at High Temperature

Single crystals of naphthalene and p-tarphenyl doped with pentacene have been polarized at 77 K/270 K and in 0.3 T, and kept polarized in lower field than 0.001 T [24]. Electrons in pentacene molecules are diamagnetic on the ground state, because electrons are on the singlet state. Electrons are excited to higher singlet states with a laser beam. The spin-orbit interaction causes the transition from singlet excited states to intermediate triplet states.

The populations of the Zeeman sublevels of the lowest triplet state with Z-components of the total electron spin +1, 0, and -1 are 12, 76 and 12 %, respectively. During the lifetime of the triplet state (~20 microsec.) microwave irradiation is performed and the external field is swept simultaneously to transfer the population difference of electrons between two Zeeman sublevels to the proton polarization by means of "the integrated solid effect". Electrons on the triplet state decay to the ground state, where the proton spin remains polarized for a long time, since the spin-spin interaction between electrons and protons on the ground state is negligible.

The proton polarization of 70 % was obtained at the liquid N_2 temperature. It is expected that the high proton polarization can be obtained at slightly lower temperature than 0°C, since the coupling between the slow molecular motion and the proton spin is small in this temperature region.

This method is applicable to wide fields of particle and nuclear physics as well as solid state physics, e.g. studies of nuclear reactions with very low momentum particles, high resolution NMR systems, structure of biopolymers.

An application of this method to scattering of unstable nuclei on the polarized proton target is presented in this WS [25].

Furthermore, it may be applicable to quantum computing, because it can be operated at high temperature and in low magnetic field.

8. Conclusion

The first experiment with the solid polarized proton target was carried out with LMN at the beginning of 1960's. Development of the solid polarized target has been continuously performed for the last 45 years. Higher polarization could be obtained in lower temperature. Most of the materials with spin can be polarized in agreement with the equal spin temperature model. The polarization could be kept for long time in the dilution refrigerator without microwaves. Targets with high dilution factors, diols, butanol, NH_3, LiD, HD etc., are useful for spin physics. The DNP of aromatic molecules in high temperature is hopeful.

Future applications of the polarized target to wide field of science are expected.

References

1. A. W. Overhauser, *Phys. Rev.*, **92**, 411 (1953)
2. A. Abragam, *Phys. Rev.*, **98**, 1729 (1955)
3. A. Abragam et al., *Phys. Lett.*, **2**, 310 (1963)
4. T. J. Schmugge, C. D. Jeffries, *Phys. Rev. Lett.*, **9**, 268 (1963)
5. A. Masaike, Memorandum of CERN-NP (1971)
6. S. Mango et al., *Nucl. Instr. Meth.*, **72**, 45 (1969)
7. H. Glättli et al., *Phys. Lett.*, **29A**, 250 (1969)
8. A. Masaike et al., *Phys. Lett.*, **30A**, 63 (1969)
9. D. Hill et al., *Phys. Rev. Lett.*, **23**, 460 (1969)
10. T. J. Schmugge, C. D. Jeffries, *Phys. Rev.*, **138A**, 1785 (1965)
11. F. M. Russell, *Proc. 2nd Int. Conf. Pol. Targets, Berkeley*, pp.89 (1971)
12. T. Niinikoski, F. Udo, *Nucl. Instr. Meth.*, **134**, 219 (1976)
13. S. Isagawa et al., *Nucl. Instr. Meth.*, **154**, 213 1978)
14. J. Deregel et al., *Proc. Int. Conf. on High Energy Physics with Pol. Beam and Targets, Lausanne*, pp.463 (1980)
15. K. Scheffler, *Proc. 2nd Int. Conf. Pol.Targets, Berkeley*, pp.271 (1971)
16. W. Meyer et al., *Nucl. Instr. Meth.*, **A215**, 65(1983)
    W. Meyer, *Nucl. Instr. Meth.*, **A526**, 12 (2004)
17. M. Gueron et al., *Phys. Rev. Lett.*, **3**, 338 (1959)
18. V. Bouffard et al., *J. Phys.*, **41**, 1447 (1980)
19. J. Ball, *Nucl. Instr. Meth.*, **A526**, 7 (2004)
20. N. Doshita, in these proceedings
21. A. Honig, *Phys. Rev. Lett.*, **19**, 1009 (1967)
22. M. Bassan et al., *Nucl. Instr. Meth.*, **A526**, 163(2004)
23. X. Wei et al., *Nucl. Instr. Meth.*, **A526**, 157 (2004)
24. M. Iinuma et al., *Phys. Rev. Lett.*, **84**, 171 (2000)
25. T. Wakui et al., in these proceedings

Atomic Beam Method

RHIC POLARIZED HYDROGEN JET TARGET, EXPERIENCE AND PROSPECTS

T. WISE*

University of Wisconsin, Madison, WI, USA
E-mail: wise@physics.wisc.edu

The polarized hydrogen jet target for RHIC has been successfully employed for measurement of A_N in the Coulomb-nuclear interference region of p-p elastic scattering at 100 GeV, for an absolute measurement of the RHIC proton beam polarization, and for calibration of the RHIC p-carbon polarimeters. Principles of operation, operating experience and experimental results are reported.

1. Overview

In a conventional atomic beam source hydrogen atoms are produced with an rf discharge inside a pyrex or quartz dissociator tube cooled by a water jacket. To match the atom velocity distribution to the optics of the system the hydrogen atoms are usually cooled with a 50–80 K tapered aluminum nozzle before exiting the dissociator through the nozzle's 2 mm diameter aperture. Thermal accommodation of atoms in the RHIC jet differs from the conventional design by the addition of a ~10 cm long extension to the glass dissociator tube between the region of intense discharge and the cold aluminum nozzle. A thermal gradient is defined over this length by clamping material of intermediate thermal conductivity (stainless steel) around the tubing extension. A good overview of the source layout can be found in Ref. [1].

Beam focusing and spin separation are accomplished by directing a collimated portion of the cooled atoms through a system of Halbach type [2] segmented Ne-Fe-B permanent magnets. Magnets of this type, well described in Ref. [3], produce pole tip fields of typically 1.6–1.7 T with a quadratic radial field dependence inside the magnet bore. Hydrogen atoms experience a force proportional to the gradient of the field with the direc-

---

*for the RHIC jet polarimeter group, see Ref. [8].

tion of the force inward for atoms with $m_f = +1/2$ (hyperfine states $|1\rangle$ and $|2\rangle$) and outward for states $|3\rangle$ and $|4\rangle$.

Design of a magnet system to produce an intense atomic beam is a surprisingly difficult problem. Even with the high pole-tip fields of the Halbach type magnets, the focusing force is relatively weak and the incident atomic beam has a broad velocity distribution making the optics problem highly chromatic. The search for an optimum design is further complicated by scattering of desired atoms off a gas halo formed by the rejected $m_f = -1/2$ atoms and by undissociated H_2 gas. Additionally there is atom-atom and atom-molecule scattering within the beam itself that depends on the focusing properties of the magnets.

2. Jet Design and Operation

The optimization problem was recently studied by Wise et al. [4,5] and the resulting design was employed during construction of the RHIC jet [1]. The magnet design is shown in Fig. 1. Figure 2 shows the fraction of atoms transmitted vs. atom velocity for that design. The success of the optimization is evident in the relatively wide velocity window with transmission ~80%.

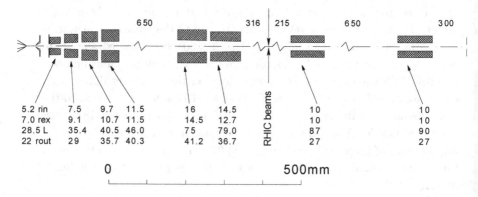

Figure 1. Sixpole permanent magnet system used with the RHIC Jet Target. Crossing with the RHIC beams occurs near a beam waist 316 mm after the sixth magnet element. Elements to the right of the RHIC crossing are part of the polarimeter system. The drawing is to scale with all dimensions in mm.

That design in conjunction with the modified dissociator system and a six stage differential pumping system produces an atomic beam with a measured [1] intensity of $12.4 \pm 0.2 \times 10^{16}$ atoms/s. The jet profile at the target plane was directly measured by scanning the entire jet assembly across the path of one of the ~1 mm diameter circulating RHIC proton beams and observing count rates in silicon detectors located near the crossing point. Figure 2 shows the results of that scan.

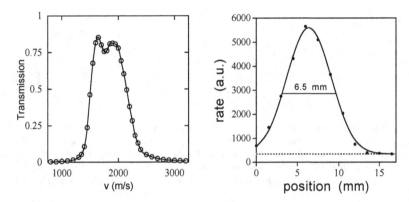

Figure 2. Left: calculated fractional transmission from magnet entrance to target region for this magnet design. Right: atomic beam profile at the target determined by scanning the Jet across one RHIC proton beam.

Figure 3 shows the energy levels and polarizations of the four hydrogen hyperfine states as a function of applied magnetic field. The beam exiting the last 6-pole separation magnet consists of nearly equal populations of hyperfine states $|1\rangle$ and $|2\rangle$. It is readily seen that to produce high polarization one needs to employ rf transitions [6,7] which exchange the populations of selected hyperfine states in the beam: either a weak field transition (WF_{1-3}) which moves state $|1\rangle$ atoms into state $|3\rangle$ giving states ($|2\rangle + |3\rangle$) and polarization $P \approx -1$, or a 2-4 strong field transition (SF_{2-4}) giving states ($|1\rangle + |4\rangle$) and $P \approx +1$. At the applied field of 0.12 T the maximum achievable two-state atom polarization is ±0.961.

Trajectory calculations for state $|1\rangle$ atoms are shown in Fig. 4. After crossing the RHIC beams at $z = 1.3$ m the atoms pass through two additional six-pole magnets and enter the polarimeter detector. A second calculation is shown in which the WF_{1-3} rf transition at $z = 1.1$ m is active. On-axis beam blocking discs ensure that 100% of state $|3\rangle$ atoms are lost to the beam on the way to the polarimeter detector at $z = 2.8$ m. The only atoms that arrive at the detector are the few state $|1\rangle$ atoms that the WF_{1-3} transition failed to flip. The same is true for states $|2\rangle$ and $|4\rangle$ atoms

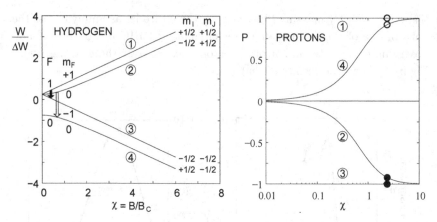

Figure 3. Left: energy level diagram for hydrogen. Energy is measured in units of ΔW, the zero field hyperfine splitting. Magnetic field, χ, is measured in units of $B_c = 50.7$ mT. Arrows indicate the effect of WF_{1-3} and SF_{2-4} rf transitions. Right: polarization vs. magnetic field for the four atomic hydrogen hyperfine states. The field at the jet target is $\chi = 2.37$ resulting in a maximum possible two-state polarization of $|P| = 0.961$.

when the SF_{2-4} transition is on. Therefore the beam leaking through the system when both rf transitions are on is a measure of the sum $\varepsilon_{1-3} + \varepsilon_{2-4}$.

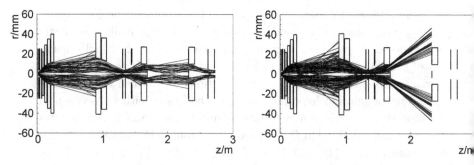

Figure 4. Left: trajectory calculations for state $|1\rangle$ atoms through the six-pole magnet system. Of the atoms arriving at the target region ($z = 1.3$ m) about 33% are transmitted to the polarimeter at $z = 2.8$ m. Right: the same calculation except that the state $|1\rangle$ atoms are assumed converted to state $|3\rangle$ by a WF_{1-3} transition located at $z = 1.1$ m.

The atom polarization delivered by the jet can be calculated from the occupation numbers of the four hyperfine states combined with the applied target field. Figure 5 shows the evolution of the beam's occupation numbers for the case when the WF_{1-3} rf transition is active: at the entrance to the first 6-pole magnet where all states are equally populated, at the exit of the last 6-pole magnet, at the target zone, and finally at the polarimeter detector. The σ_i are optical transmission factors for the 6-pole magnets. Optical tracking codes indicate that σ_3 and σ_4 are both rather small ($\leq 10^{-3}$) implying nearly complete rejection of those atoms by the 6-pole magnets. The coefficient ε_{1-3} (ε_{2-4}) represents the small fraction of state $|1\rangle$ (state $|2\rangle$) atoms the WF$_{1-3}$ (SF$_{2-4}$) transition fails to flip into state $|3\rangle$ (state $|4\rangle$). The presence of on-axis beam blocking discs in the polarimeter section (see Fig. 4) ensures that σ'_3 and σ'_4 are identically zero. When properly tuned, typical values for ε_{1-3} and ε_{2-4} are 0.003 or less. Similarly, the deviation of σ_1/σ_2 from 1 is about 1%. We routinely measure $P_{\text{atom}} = 0.958 \pm 0.001$.

$$\begin{pmatrix} N \\ N \\ N \\ N \end{pmatrix} \Rightarrow \text{6-poles} \begin{pmatrix} N\sigma_1 \\ N\sigma_2 \\ 0 \\ 0 \end{pmatrix} \Rightarrow \text{WF}_{1-3} \begin{pmatrix} \varepsilon_{1-3}N\sigma_1 \\ N\sigma_2 \\ (1-\varepsilon_{1-3})N\sigma_1 \\ 0 \end{pmatrix} \Rightarrow \text{polarimeter} \begin{pmatrix} \varepsilon_{1-3}N\sigma'_1 \\ N\sigma'_2 \\ 0 \\ 0 \end{pmatrix}$$

Figure 5. Evolution of the four hydrogen hyperfine occupation numbers for the case of a WF$_{1-3}$ rf transition located immediately in front of the target zone.

An unpolarized molecular component to the beam is also present. That component has been measured in two ways, with a quadrupole mass spectrometer, and by ionization with a 600 eV electron beam. Taking that dilution into account the target polarization is $P = 0.934 \pm 0.008$ where the error is entirely dominated by the uncertainty in the molecular dilution.

Figure 6 shows typical polarimeter data taken during the normal polarization cycling of the jet. When the polarization state is changed there is an intentional overlap period when both the WF$_{1-3}$ and the SF$_{2-4}$ rf transitions are on. Thus $\varepsilon_{1-3} + \varepsilon_{2-4}$ is continuously monitored during data taking.

3. Results

In 2004, the jet was installed into RHIC as an internal target at the 12 O'clock region. The target chamber was instrumented with an array of Si strip detectors near 90° in the lab enabling a measurement of A_N at low momentum transfer for the p-p elastic channel. Our result is shown in Fig. 7. A complete description of the measurement has been given in Ref. [8]. We also used the jet for an absolute measurement of the RHIC beam polar-

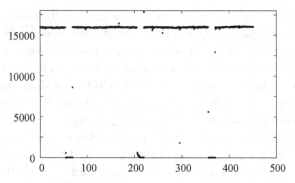

Figure 6. Typical data taking cycle for the RHIC jet. Beam intensity at the polarimeter detector is plotted vs. time. At $t = 0$ the SF_{2-4} transition is on and the WF_{1-3} is off. The detected beam is around 16,000 Hz and the target polarization is $P \approx -1$. At $t = 50$ the WF_{1-3} is turned on and the detected rate drops to 35 Hz. At $t = 65$ the SF_{2-4} is turned off, the rate rises to $\sim$16,000 Hz and the polarization is $P \approx +1$. At $t = 200$ the SF_{2-4} is turned on. In this instance the SF_{2-4} transition has turned on slowly due to a delay in its 1.43 GHz phase sensitive loop but the rate eventually drops to 35 Hz confirming proper operation of both transitions. At $t = 215$ the WF_{1-3} is turned off completing the cycle.

ization. We compare $\varepsilon_{\text{beam}}$, the L-R asymmetry with the polarized beam, to $\varepsilon_{\text{target}}$, the asymmetry with the polarized jet target, using the relation $P_{\text{beam}}/P_{\text{jet}} = -\varepsilon_{\text{beam}}/\varepsilon_{\text{jet}}$ which takes advantage of $A_N^{\text{beam}} = -A_N^{\text{target}}$ in p-p elastic scattering.

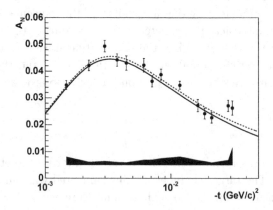

Figure 7. Results for A_N in p-p elastic scattering at 100 GeV/c incident beam and low momentum transfer. The curves are model calculations without (solid) and with (dotted) a hadronic spin flip term [9].

Because the jet is relatively thin it is only possible to determine the average RHIC beam polarization over a number of fills. For the 2004 run, the average circulating beam polarization was measured to be $P_{beam} = 0.392 \pm 0.026$. The above data can also be used for an absolute calibration of fast p-carbon polarimeters located elsewhere in the ring.

Design of an additional rf transition that will allow production of a vector and tensor polarized deuterium jet is in progress.

References

1. A. Zelenski et al., *Nucl. Instr. and Meth.* **A536**, 248 (2005).
2. K. Halbach, *Nucl. Instr. and Meth.* **A169**, 1 (1980).
3. A. Vassiliev et al., *Rev. Sci. Inst.* **71**, 3331 (2000).
4. T. Wise et al., *Nucl. Instr. and Meth.*, **A556**, 1 (2006).
5. T. Wise et al., in *15th International Spin Physics Symposium*, Y. I. Makdisi, A. U. Luccio, and W. W. MacKay Eds., *AIP Conf. Proc.* **675**, 934 (2003).
6. W. Haeberli, *Ann. Rev. Nucl. Sci.* **17**, 373 (1967).
7. A. Airapetian et al., *Nucl. Instr. and Meth.* **A540**, 68 (2005).
8. H. Okada, I. G. Alekseev, A. Bravar, G. Bunce, S. Dhawan, R. Gill, W. Haeberli, O. Jinnouchi, A. Khodinov, Y. Makdisi, A. Nass, N. Saito, E. J. Stephenson, D. N. Svirida, T. Wise, and A. Zelenski, submitted to *Phys. Rev. Lett.*
9. N. H. Buttimore et al., *Phys. Rev.* **D59**, 114010 (1999).

POLARIZED INTERNAL GAS TARGET IN A STRONG TOROIDAL MAGNETIC FIELD

E.TSENTALOVICH, E. IHLOFF, H.KOLSTER, N.MEITANIS,
R.MILNER, A.SHINOZAKI, V.ZISKIN, Y.XIAO, C.ZHANG

MIT-Bates Linear Accelerator Center,
Laboratory for Nuclear Science,
21 Manning Rd, Middleton, MA, USA 01949

A polarized hydrogen/deuterium internal gas target has been constructed and operated at the internal target region of the South Hall Ring (SHR) of the MIT-Bates Linear Accelerator Center to carry out measurements of spin-dependent electron scattering at 850 MeV. The target used an Atomic Beam Source (ABS) to direct a flux of highly polarized atoms into a thin-walled, coated storage cell. The polarization of the electron beam was determined using a Compton laser backscattering polarimeter. The target polarization was determined using well known nuclear reactions. The ABS and storage cell were embedded in the Bates Large Acceptance Toroidal Spectrometer (BLAST), which was used to detect scattered particles from the electron-target interactions. This target was the first to be operated inside a magnetic spectrometer in the presence of a magnetic field exceeding 2 kG. An ABS intensity $2.5 \cdot 10^{16} at/sec$ and a high polarization ($\approx 70\%$) inside the storage cell have been achieved.

1. Introduction

The study of the structure of the nucleon and light nuclei with spin-dependent electron scattering continues to be a subject of intense experimental and theoretical interest. The BLAST experiment [1] at the MIT-Bates Linear Accelerator Center was designed to measure in a comprehensive way spin-dependent electron scattering from the proton and deuteron at low momentum transfer, i.e. $q^2 \leq 0.8 (GeV/c)^2$. BLAST uses a stored beam of longitudinally polarized electrons incident on a polarized internal gas target in the South Hall Ring (SHR) [2]. This technique permits essentially background-free measurements with little or no dilution of the experimental signal from unpolarized nucleons in the target. In addition, the extremely low-mass target-interaction area facilitates clean detection of final state products (protons, neutrons, deuterons, pions etc.) in coincidence with the scattered electron.

To carry out the BLAST scientific program, a polarized hydrogen and deuterium target was constructed and installed in early 2003. The target was commissioned in that year and production data were taken from November 2003 to May 2005.

2. Experimental setup

The MIT-Bates accelerator complex includes a polarized electron source, linear accelerator with a recirculator and 1 GeV storage ring. The polarized source [3] uses a strained GaAs cathode with high-gradient doping. The electrons are accelerated in the accelerator-recirculator complex to 850 MeV and then injected into the SHR. The injection takes about 30 sec, and the injected current is up to 225 mA. The Siberian Snake built in the Budker Institute for Nuclear Physics (Novosibirsk) [4] was installed in the ring. It keeps the longitudinal beam polarization in the internal target area at an average polarization of 67%. The beam polarization was constantly monitored by a Compton polarimeter [5]. The BLAST (Bates Large Acceptance Spectrometer Toroid) magnetic spectrometer was used to detect scattered electrons and secondary particles in coincidence. BLAST is an open geometry detector with a toroidal magnetic field.

The ABS had to be located at the center of the BLAST magnetic toroid within an approximately cylindrical volume of about 1 meter in diameter. Thus, the dissociator, focusing sextupole magnets, vacuum pumping and diagnostic instrumentation were located in a region of high magnetic field. The BLAST magnetic field was almost zero along the axis of the toroid (storage cell location) and it increased gradually to reach the maximum of about 2.2 kG at the location of the first sextupole magnet and then slowly decreased.

The ABS produced a jet of polarized atoms and injects it into the T-shaped storage cell (60 cm length, 15 mm diameter) cooled to about 100 K. A small outlet allowed some fraction of the jet to pass through the cell for polarization analysis in the Breit-Rabi Polarimeter (BRP). The holding field magnet produced the magnetic field $B_{hold} \sim 500\,G$ in the storage cell area that defines the orientation of target polarization. During the experiment, scattering from the target gas usually reduced the stored beam lifetime by 10-15%.

3. The Atomic Beam Source

The RF discharge in the dissociator separated hydrogen (or deuterium) molecules into atoms, and an atomic jet was formed in the nozzle. A 4.5 mm diameter skimmer was located 12 mm below the nozzle, and a 7.5 mm diameter aperture separated the skimmer chamber from the sextupole chamber.

Two sets of sextupole magnets were used to focus atoms from the upper hyperfine states into the entrance of the storage cell and defocus atoms from the lower states. A Medium Field Transition unit (MFT) was located between the first and the second sextupoles, while Strong and Weak Field Transition units (SFT and WFT) were located below the second sextupole.

The RF coil covered only a fraction of the dissociator tube, leaving a significant area between the coil and the nozzle uncovered. However, it was found that the discharge had to extend into this area almost to the nozzle for the degree of dissociation to be high. In order to control the discharge area, two optical sensors have been placed at the pyrex tube: one at the middle of the RF coil, and one close to the nozzle. It was found that the BLAST magnetic field does not affect the discharge inside the coil, but it did extinguish the discharge near the nozzle, and the degree of dissociation was reduced. Magnetic shielding of the dissociator eliminated this effect.

The vacuum pumping speed is a major factor affecting ABS intensity. At a higher gas flow the scattering from the residual gas in the ABS actually decreases the atomic beam intensity.

There are 4 different vacuum chambers in the ABS: nozzle chamber, skimmer chamber, top sextupole and lower sextupole chambers. At the early stage of the ABS commissioning, the effects of the jet scattering on the residual gas were investigated by leaking additional hydrogen gas into different chambers of the ABS. The results allowed an identification of the areas where the pumping speed improvement would make the largest impact in maximizing the ABS intensity (nozzle chamber and bottom sextupole chamber). In the final version of the target the pumping speed was increased in the first two vacuum chambers, the NEG pumps have been replaced with cryopumps in the last two, and vacuum conductivity was improved in the SFT unit.

Two sets of permanent sextupole magnets were used in the ABS. The location and profile of the magnets has been chosen according to the results of Monte-Carlo simulations in order to maximize the transmission of the atoms with electron spin $+1/2$ and to minimize it for the atoms with spin

-1/2.

The presence of the BLAST magnetic field was observed to produce a significant effect on the focusing in the sextupoles. Although the BLAST field was rather uniform, and it did not change the amplitude or direction of the gradients in sextupole magnets, it changed the direction of the magnetic moment of the atoms relative to the direction of the gradient. Generally, the force acting on the magnetic dipole is $\vec{F} = \vec{\nabla} \cdot (\vec{\mu} \cdot \vec{B})$, and since dipoles follow the direction of the magnetic field $\vec{\mu} = \mu \cdot \vec{B}/B$, it could be transformed to $\vec{F} = \mu \cdot \vec{\nabla} B$. In an ideal sextupole ($B_x = G(x^2 - y^2)$; $B_y = -G \cdot 2xy$) with no external field the force has only a radial component:

$$\vec{F} = 2G\mu\vec{r}; \quad \frac{\vec{F}}{F} \cdot \frac{\vec{r}}{r} = 1$$

An external magnetic field B_0 applied in the x-direction (so now $B_x = G(x^2 - y^2) + B_0$) does not affect the amplitude of the force, but it does change it's direction:

$$\frac{\vec{F}}{F} \cdot \frac{\vec{r}}{r} = \frac{1 + b \cdot \cos(2\theta)}{\sqrt{1 + b^2 + 2b \cdot \cos(2\theta)}},$$

where b is a ratio of external field and sextupole field at the given point $b = B_0/B_6$ and θ is a polar angle. One can see that in the regions where the external field exceeds the sextupole field, the y-component of the force changes sign and becomes defocusing! Since the BLAST field strength was over 2 kG, and the sextupole field vanishes at the center of the sextupoles, and was about 10 kG at the pole tip, the effect was very significant and reduced the ABS intensity by factor of 2. To minimize the effect the sextupoles were encased in magnetic shields. The ABS intensity loss was reduced to less than 10%.

4. Breit-Rabi Polarimeter

The convenient BRP with a permanent sextupole magnet, a chopping wheel and quadrupole mass analyzer (QMA) could not be used in the BLAST environment. The QMA does not work in a strong magnetic field, unless it is placed at the very bottom of the BLAST pit, some 2 m away from the target. At this distance the signal was too weak for reliable measurements. The only natural way to enhance the signal, the compression tube, has too slow a response time to be combined with a chopping wheel.

Instead, a BRP with a dipole magnet was used (Fig. 1). The magnet had a very strong (about 2.5 kG/cm) and uniform gradient and was placed after a small (2 mm) diaphragm below the outlet of the storage cell. Three compression tubes (CT) had been installed 1.5 m below the magnet. The compression tubes greatly enhanced the signal from the ballistic particles.

The flux of $6 \cdot 10^{13} at/sec$ into the tube 75 mm long and 5 mm diameter produced a signal of about $1 \cdot 10^{-6} torr$ in the vacuum gauge. With a total volume of the tube and vacuum gauge of the order of 1 liter the typical response time was about 1 sec.

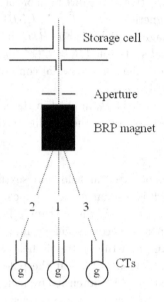

Figure 1. BRP layout. CTs- Compression Tubes equipped with vacuum gauges (Left, Central and Right). 1 - trajectories of molecules, 2 - trajectories of atoms with electron spin of +1/2, 3 - trajectories of atoms with spin -1/2.

With the dipole magnet turned off, the central CT collected both atoms and molecules. With the magnet on, the atoms were deflected into the left or right CT depending on their electron polarization. The BRP allowed measurement of both the degree of dissociation in the discharge tube (ABS sextupole magnets have been moved out for these measurements) and the polarization in the atomic beam. Moreover, one could monitor the polarization of the atomic beam during the production run observing the signals from the left and right CTs. The drawback of the system is that only the central trajectories of the atomic beam are sampled in the BRP, and therefore the absolute measurements of polarization contain significant systematic uncertainties.

5. Target Performance

Target intensity and polarization were monitored on a daily basis by measuring the rates and asymmetries of elastic and inelastic scattering. The ABS flow into the cell for both hydrogen (1 state) and deuterium (2 states) was about $2.5 \cdot 10^{16} at/sec$, which produced target thickness of $7 \cdot 10^{13} at/cm^2$. The measured polarization of the deuterium target, avereged over several months of running, was Pz≈86%, Pzz≈68%. Polarization of the hydrogen target was Pz≈82%.

References

1. R.Alarcon et al., Nucl. Phys. A 663 (2000), p.1111.
2. W.A.Franklin et al., SPIN2004 Conf. Proc. World Scientific, 2004.
3. M.Farkhondeh et al., Proc. of 15th International Spin Physics Symposium, Danvers, 2002, AIP 675, p.1098.
4. T. Zwart et al., IEEE Proc. PAC (1995), p.600.
5. W.A.Franklin et al., SPIN2002 Conf. Proc. AIP 675, p.1058.

AN ATOMIC BEAMLINE TO MEASURE THE GROUND-STATE HYPERFINE SPLITTING OF ANTIHYDROGEN

B. JUHÁSZ,* E. WIDMANN

Stefan Meyer Institut für subatomare Physik,
Boltzmanngasse 3, A-1090 Vienna, Austria

D. BARNA,† J. EADES, R.S. HAYANO, M. HORI, W. PIRKL

Department of Physics, University of Tokyo,
7-3-1 Hongo, Bunkyo-ku, Tokyo 113-0033, Japan

D. HORVÁTH

KFKI Research Institute for Particle and Nuclear Physics,
H-1525 Budapest, Hungary
and
Institute of Nuclear Research of the Hungarian Academy of Sciences,
H-4001 Debrecen, Hungary

T. YAMAZAKI

DRI, RIKEN, Wako, Saitama 351-0198, Japan

The ASACUSA collaboration at CERN-AD is designing an atomic beamline to measure the hyperfine splitting of the ground state of antihydrogen. The apparatus will consist of two sextupoles for spin selection and analysis, and a microwave cavity to flip the spin. The beamline has to have a high transmission efficiency, which requires an unusual design. Numerical simulations show that the splitting frequency can be measured with an accuracy of 10^{-6} or better.

---

*On leave from the Institute of Nuclear Research of the Hungarian Academy of Sciences, H-4001 Debrecen, Hungary

†On leave from KFKI Research Institute for Particle and Nuclear Physics, H-1525 Budapest, Hungary

1. Motivation of the experiment

The antimatter counterpart of the simplest neutral atom, the antihydrogen ($\overline{\text{H}}$), which consists of an antiproton ($\overline{\text{p}}$) and a positron, can be used to test the Charge-Parity-Time reversal (CPT) invariance. The 1s ground state of hydrogen (antihydrogen) is split due to the interaction between the electron (positron) spin and the proton (antiproton) spin, and can have quantum numbers $F = 0, 1$ (total spin) and $M = -1, 0, 1$ (projection of F). The states with quantum numbers $F = 0$ (singlet state) and $F = 1$ (triplet state) have different energies even at zero external magnetic field.

The splitting frequency in hydrogen (antihydrogen) is proportional to both the electron (positron) and proton (antiproton) spin magnetic moments. The antiproton spin magnetic moment $\mu_{\overline{\text{p}}}$ is known to a precision of only 0.3%, therefore a measurement of ν_{HF} for antihydrogen (which has never been done before) with a precision of only 10^{-6} can already lead to an improvement of the value of $\mu_{\overline{\text{p}}}$ by three orders of magnitude.

In the recent years, the group of V.A. Kostelecky has developed an extension of the standard model of elementary particles that includes both CPT violating and Lorentz invariance violating terms in the Lagrangian of the quantum field theory [1]. The parameters introduced by this theory have a dimension of energy (or frequency) i.e. they have an absolute magnitude rather than relative difference. Thus this theory claims that it is not the relative but the *absolute* precision of an experiment that matters, therefore the $\overline{\text{H}}$ GS-HFS measurement can be competitive with the oft-quoted 'most sensitive CPT limit' of $|m_{K^0} - m_{\overline{K}^0}|/m_{K^0} < 10^{-18}$ [2].

2. Proposed experimental method

The ASACUSA collaboration at CERN-AD plans to use a method [3] similar to classical atomic beam (Stern-Gerlach type) experiments to measure the $\overline{\text{H}}$ GS-HFS. A schematic view of the proposed spectrometer line is shown in Fig. 1.

The antihydrogen atoms will be produced in either a two-frequency Paul trap or a cusp trap, which are currently under development [3]. Depending on the sign of their magnetic moment, the four hyperfine states of the ground-state $\overline{\text{H}}$ atoms can be divided into two pairs (see Fig. 2): 'high-field seekers', which will move towards regions of higher magnetic field in the inhomogeneous field of a sextupole magnet and thus will be defocused, and 'low-field seekers', which will be attracted towards the sextuple axis and thus will be focused onto a microwave cavity. The focused atoms will

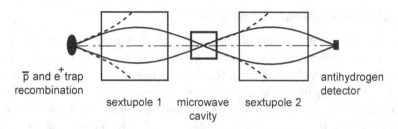

Figure 1. Schematic view of the proposed atomic beam spectrometer with the antihydrogen source, the two sextupole magnets and the microwave cavity. The trajectories drawn with solid lines represent $\overline{H}$ atoms in low-field seeker states, while the dashed lines represent $\overline{H}$ atoms in high-field seeker states.

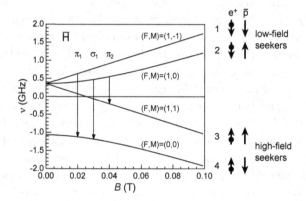

Figure 2. Frequencies (i.e. energies) of the four hyperfine states of antihydrogen as a function of the external magnetic field B. The transitions observable with the proposed method are also drawn.

then pass through a second sextupole, which will again focus them onto an antihydrogen detector. However, if a $\overline{H}$ atom in e.g. the $(F, M) = (1, -1)$ low-field seeker state is converted into the $(0, 0)$ high-field seeker state by a microwave field of appropriate frequency in the cavity, then the atom will be defocused and will not reach the detector. Thus the on-resonance count rate in the detector will drop from the constant off-resonance count rate.

Antihydrogen, unlike hydrogen, is very precious, therefore the spectrometer must have high transmission efficiency. This can be reached by an unusually large acceptance angle ($\sim 20°$), no collimators or skimmers, and a high magnetic field. The satisfy the latter requirement, the usage of superconducting magnets is preferred, which can have a pole tip field of several Tesla. However, this strong field has to be carefully shielded away

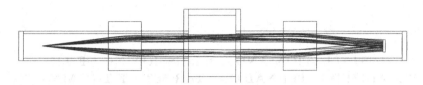

Figure 3. Simulated trajectories of the antihydrogen atoms (with microwave resonance off) in the spectrometer line. The total length from the source to the detector is ∼150 cm.

from the microwave cavity, otherwise the magnetic field would shift the resonance lines considerably. Moreover, the spin of the atoms have to be flipped using a non-adiabatic π-pulse in the cavity. This is in contrast to the adiabatic transitions commonly used in atomic beam sources.

3. Monte Carlo simulations

Simulations using the GEANT4 toolkit [4] have been carried out to estimate the expected count rates and experimental resolution. The preliminary results showed that a spectrometer line with sextupoles, each with an internal diameter of 10 cm, an effective length of 12 cm, and a pole tip field of 4 Tesla (see Fig. 3), can have a total transmission efficiency of 5–20 $\times 10^{-5}$. Using the expected production rate of 200 $\overline{\text{H}}$/s, the detection rate at the antihydrogen detector will be 0.5–2 $\overline{\text{H}}$/min. The expected resonance width is a few kHz, while the center of the resonance can be determined with a precision below 1 kHz, which corresponds to a relative precision of $< 10^{-6}$.

Acknowledgements

This work was supported by the Grant-in-Aid for Specially Promoted Research (15002005) of Monbukagakusho of Japan, the exchange visit programme between the Austrian and the Hungarian academies of science, and the Hungarian Scientific Research Fund (OTKA T046095).

References

1. R. Bluhm, V.A. Kostelecky, and N. Russell, *Phys. Rev. Lett.* **82**, 2254 (1999).
2. S. Eidelman *et al.*, *Phys. Lett.* **B592**, 1 (2004).
3. ASACUSA proposal addendum, CERN/SPSC 2005-002, SPSC P-307 Add.1 (2005).
4. S. Agostinelli *et al.*, *Nucl. Instrum. Methods* **A506**, 250 (2003).

INSTALLATION AND COMMISSIONING OF THE POLARIZED INTERNAL GAS TARGET OF THE MAGNET SPECTROMETER ANKE AT COSY-JÜLICH

R. ENGELS, D. CHILADZE*, S. DYMOV**, K. GRIGORYEV***,
D. GUSEV**, B. LORENTZ, D. PRASUHN, F. RATHMANN, J. SARKADI,
H. SEYFARTH, AND H. STRÖHER

Institut für Kernphysik, Forschungszentrum Jülich, 52425 Jülich, Germany
* *PhD student from Tbilisi State University, 380662 Tbilisi, Georgia*
** *PhD student from Joint Inst. for Nuclear Research, 141980 Dubna, Russia*
*** *PhD student from Petersburg Nucl. Phys. Inst., 188300 Gatchina, Russia*
E-mail: r.w.engels@fz-juelich.de

S. MIKIRTYCHYANTS, M. MIKIRTYCHYANTS, AND A. VASILYEV

Petersburg Nuclear Physics Institute, 188300 Gatchina, Russia

For future few-nucleon interaction studies with polarized beams and targets at COSY (Jülich, Germany), a polarized internal storage-cell gas target has been developed. Polarization measurements are performed with a Lamb-shift polarimeter. For commissioning and first experimental studies, the polarized target has been installed at the magnet spectrometer ANKE. COSY-beam properties, achieved with stacking injection and electron cooling, have been investigated. The vertex distribution at the target is investigated by events from the $pp \to d\pi^+$ reaction. Results of these studies are presented.

1. Introduction

The polarized proton and deuteron beams of the storage ring COSY (Jülich, Germany) and the polarized internal target (PIT) [1], being installed at the magnet spectrometer ANKE, will allow essential double-polarization experiments [2]. The storage cell of the PIT is fed by the $\vec{H}$ or $\vec{D}$ beam from the polarized atomic beam source (ABS) [3]. The $\vec{H}$-beam intensity in two hyperfine states, measured with a compression-tube setup, amounts to $(7.4 \pm 0.3) \times 10^{16}$ atoms/s [1]. Preliminary measurements indicate a similar beam intensity for the $\vec{D}$ beam in three hyperfine states. At the entrance of the feeding tube of the storage cell, the beam width (FWHM) was measured as (6.9 ± 1.0) mm [3]. With the Lamb-shift polarimeter (LSP) [4,5], developed at Universität zu Köln, the vector polarization of the $\vec{H}$ beam was measured as $+0.898 \pm 0.004$ [5] and -0.960 ± 0.005 [6]. Preliminary values for the vector and tensor polarization of the $\vec{D}$ beam have been measured as well. Final values, however, can only be given after final fine-tuning of the hyperfine transition units.

In order to study the lateral dimensions of the COSY beam at the ANKE-target position and its intensity with the storage cell at the target position, a setup has been developed which allows one to position diaphragms and storage cells onto the beam axis, to move them perpendicular to the beam axis in horizontal ($\Delta X = \pm 75$ mm) and vertical ($\Delta Y = \pm 12.5$ mm) direction, and to tilt the cell-tube axis against the COSY beam axis (Fig.1). The remote-control system is part of the PIT control and interlock installation [7].

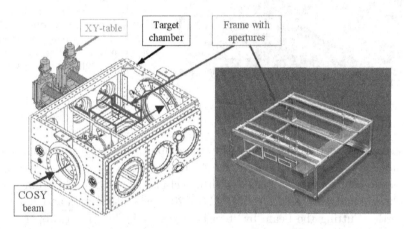

Figure 1. The new ANKE target chamber (width 60 cm, length 80 cm, and height 40 cm) with the motorized, remote-controlled XY manipulators, used to position the target frame (left-hand part). Various diaphragms or storage cells can be suspended in the frame (right-hand part).

2. The PIT at the target position of ANKE

In June 2005, the ABS and the LSP were mounted at the ANKE-target chamber for commissioning and test measurements (Fig.2). The supply units are mounted on a movable platform, positioned near the ANKE target place inside the COSY ring area. The PIT construction and setup provides a fast exchange to a cluster target source within a couple of days.

Figure 2. The ABS (vertically above the ANKE target chamber) and the horizontally mounted LSP (lower right-hand side) in the COSY tunnel. The COSY beam circulates through the setup from the left hand side. Discernible ABS components are: at the top the upper components of the dissociator, the turbomolecular pumps, and the diaphragm pumps, mounted on the bridge which carries the ABS and the target chamber. Among the LSP components one discerns the ionizer (horizontally at the target chamber behind a gate valve), which is used to ionize polarized atoms effusing from the target cell through a sample tube into the LSP.

3. Commissioning studies

Based on earlier studies, using different diaphragms to determine the lateral COSY-beam dimensions at the ANKE-target position, more extensive measurements were performed with a rectangular $40_{\text{hor}} \times 23_{\text{vert}}$ mm^2 diaphragm and a storage-cell tube of 20×20 mm^2 cross section and 400 mm length. Cutting the beam by stepwise moving the left, right, upper, and lower edges of the diaphragm frame (Fig.1) towards the beam axis gave the lateral and vertical width of the proton beam at injection energy of 40 MeV of 16 mm horizontal and 15 mm vertical. At its centered position, the diaphragm reduces the number of stored protons to $\sim 87\%$. Application of electron cooling [8,9] results in a reduction to ~ 13 mm horizontal and ~ 8 mm vertical. The results, achieved with the storage cell, with stacking injection [8,9,] and with electron cooling are collected in Tab. 1. The number of $6.4 \cdot 10^9$ stored and accelerated protons yields an appreciable luminosity of 10^{30} cm^{-2}s^{-1} for double polarization experiments. Further studies will concern the additional application of stochastic cooling. The evaluation of the data from the $p\vec{p} \to d\pi^+$ reaction, measured with injection of $\vec{\text{H}}$ into the storage cell, is in progress.

Table 1. The number of stored protons N_p at injection energy and after acceleration to 600 MeV without and with the empty cell tube, and with the cell tube fed by gas for the undeflected beam through ANKE (α=0°) and a chikane angle of α=9.2° with stacking injection (s) and electron cooling at injection (c).

α	beam	N_p at injection			N_p at 600 MeV		
		no cell	empty cell	filled cell	no cell	empty cell	filled cell
0°	c	$8.3 \cdot 10^{10}$	$6.6 \cdot 10^9$		$1.4 \cdot 10^{10}$	$3.5 \cdot 10^9$	
0°	s+c				$2.6 \cdot 10^{10}$	$2.0 \cdot 10^9$	
9.2°	s+c		$8.5 \cdot 10^9$	$8.8 \cdot 10^9$		$6.0 \cdot 10^9$	$6.4 \cdot 10^9$

Acknowledgements

The authors want to acknowledge the important help by the infrastructure divisions of FZJ and IKP in the development and installation of the PIT. Thanks go to members of the ANKE collaboration, who enabled the installation and the commissioning studies. The authors thank E. Steffens, Universität Erlangen-Nürnberg, and H. Paetz gen. Schieck, Universität zu Köln, for their contributions to the development of the PIT.

References

1. F. Rathmann et al., Proc. 15th Int. Spin Physics Symposium, Upton, NY, USA, 2002. Y. Makdisi, A.U. Luccio, W.W. MacKay (Eds.), AIP Conf. Proc. **675** (2003) 553.
2. A. Kacharava, F. Rathmann, and C. Wilkin: Spin Physics from COSY to Fair (COSY Experiment Proposal No. 152), nucl-ex/0511028 (2005).
3. M. Mikirtytchiants et al., Proc. 9th Int. Workshop on Polarized Sources and Targets, Nashville, IN, USA, 2001. V.P. Derenchuk and B. von Przewoski (Eds.), World Scientific (2002) 47.
4. R. Engels et al., Rev. Sci. Instrum. **74** (2003) 4607.
5. R. Engels et al., Rev. Sci. Instrum. **76** (2005) 053305.
6. R. Engels et al., Proc. 6th Int. Conf. on Nuclear Physics at Storage Rings, Jülich-Bonn, Germany, 2005. D. Chiladze, A. Kacharava, and H. Ströher (Eds.), FZ Jülich, Matter and Materials **30** (2005) 381.
7. H. Kleines et al., Nucl. Instr. and Meth. A (2006, in print).
8. R. Maier et al., IKP+COSY Annual Report 1999. Report Jül-3744 (2000) 147.
9. H.J. Stein et al., Proc. 18th Conf. on Charged Particle Accelerators,(RUPAC 2002), Obninsk, Russia, 2002. I.N. Meshkov (Ed. in chief), NRCRF Obninsk (2004) 220.

Cryogenic Method

PROGRESS IN DYNAMICALLY POLARIZED SOLID TARGETS *

J. HECKMANN, S. GOERTZ,[†] CHR. HESS, W. MEYER, E. RADTKE, AND G. REICHERZ

Ruhr-Universitaet Bochum
Inst. f. Experimentalphysik AG I
Universtitaetsstrasse 150
D-44801 Bochum, Germany

Polarized solid targets have been used in nuclear and particle physics experiments since the early 1960s, and with the development of superconducting magnets and ^{3}He/^{4}He dilution refrigerators in the early 1970s, proton polarization values of 80 – 100 % have been achieved routinely in various target materials at two standard magnetic field and temperature conditions ($2.5\,T, < 0.3\,K$ and $5\,T, 1\,K$). Due to the much lower magnetic moment of the deuteron compared with that of the proton, deuteron polarization values have been considerably lower, typically 30 – 40 %. Now, however, research at the University of Bochum is yielding materials with deuteron polarizations as high as 80 %.

1. Introduction

A polarized solid target can be assumed to be an ensemble of particles with spin, placed in a high magnetic field and cooled to very low temperature. The basic idea - to obtain a high polarization of nuclear spins - consists in using a microwave field in order to transfer the polarization of electron spins to these nuclei. This process is called dynamic nuclear polarization (DNP).

The main practical problem with DNP is finding a suitable combination of hydrogen - or deuterium - rich material and a paramagnetic dopant, i. e. a material with an unpaired or quasi-free electron. Suitable means that the relaxation time of the electron spins is short ($\sim ms$) and that of the nucleons

---
*This work is supported by BMB+F.
[†]now at Bonn University

(nuclei) is long ($\sim min$), resulting in a high nucleon (nuclei) polarization. A comprehensive theoretical treatment of the DNP mechanism can be found in Refs. 1 and 2.

A very important feature of polarized solid targets is the fact, that the DNP scheme works for any nucleus with spin. Because all the target materials used are diamagnetic compounds, some amount of paramagnetic impurities (radicals or crystalline defects with unpaired electron spins) have to be implanted into the host material (doping). In the case of materials, which are liquid at room temperature (like butanol, which is one of the standard materials), this can be done by chemical doping with suited radicals. A generally applicable method is to induce paramagnetic defect electrons by irradiating the material with ionizing radiation, e.g. an intense electron beam. For an effective DNP process, a number of $2 \cdot 10^{19} spins/cm^3$ has shown to be most suitable.

Naturally, these electrons can be studied in Electron Spin Resonance (ESR) experiments, and since the properties of the paramagnetic centers strongly affect the efficiency of the DNP process, ESR spectroscopy has become an important tool in the research of polarized target materials. According to the spin temperature theory, a narrow electron spin resonance (ESR) line enables the creation of high inverse spin temperatures - and thus of high nuclear polarizations.

In the following, the spin temperature theory, which gives the connection between ESR linewidth and polarization behaviour, as well as the dominant line broadening mechanisms, will be introduced briefly. While the detailed ESR measurements of various target materials in X-band and V-band are presented in Refs. 3 and 4 respectively, the results in connection with the spin temperature theory are discussed at the end of the article.

2. The spin temperature picture of the DNP

The simplest picture to describe the DNP process is the so-called Solid State Effect (SSE), which is treated in Refs. 5 and 6, but can only be applied to some of the proton target materials, where the nuclear (proton) Larmor frequency exceeds the ESR linewidth significantly (for example LMN, one of the first used polarized solid target materials in the early 1960s).

Especially for deuterated target materials the situation is completely different: the magnetic moment of the deuteron is 6.5 times smaller than that of the proton, so that the nuclear (deuteron) Larmor frequency is usually even significantly smaller than the ESR linewidth. The DNP in these

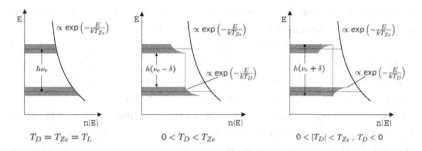

Figure 1. Distribution of the population numbers of the electronic Zeeman bands. Left: thermal equilibrium, middle: cooling, right: heating of the electronic spin-spin reservoir.

materials is well described by the spin-temperature theory (based on the Provotorov theory, which describes the magnetic resonance of spin systems in solids), which was developed by Abragam [1] and Goldman [2] and recently surveyed in Ref. 7. The spin-temperature theory treats the different spin systems as thermodynamical ensembles. In this picture an energy reservoir is assigned to each occurring interaction, characterized by a corresponding temperature. Namely, these are the electronic and nuclear Zeeman interaction and the electronic dipolar or spin-spin interaction. This dipolar interaction between the spins of the paramagnetic electrons causes a splitting of the electronic Zeeman levels into quasi-continuous bands, whose energy width is given by the so-called dipolar width D. The population numbers of the states in the different reservoirs are described by the corresponding Boltzmann distributions with the temperatures T_{Ze}, T_N and T_D, respectively. In thermal equilibrium these are all equal to the lattice temperature T_L (Fig. 1, left).

The DNP process in this thermodynamical scheme can be depicted as follows: by applying a saturating microwave field with a frequency $\nu_e \mp \delta$, where ν_e is the electronic Larmor frequency and δ is the frequency width of the Zeeman bands ($h\delta = g_e \mu_B D$), the population numbers of the inner (outer) parts of the Zeeman bands are equaled. Consequently, the distribution of the population numbers inside the bands is represented by a Boltzmann factor with a lower (or even negative) dipolar temperature T_D (Fig. 1, middle and right). Thus, the electronic dipolar reservoir can be cooled or heated, which is the case for negative dipolar temperatures. For an effective cooling (heating) of the dipolar reservoir the dipolar width D should be minimized: for a certain difference in the population numbers of the upper and lower edge of the Zeeman bands (caused by the microwave irradiation), a smaller dipolar width means a more flat Boltzmann distribution, according to a smaller absolute value of T_D.

Once the electronic dipolar reservoir is cooled (or heated) to a small absolute value of T_D, the nucleons can be polarized effectively by equalizing the nuclear Zeeman temperature T_N to T_D in terms of a thermal contact between both reservoirs. For instance, this contact can be realized by means of the hyperfine interaction, which couples the paramagnetic electrons to the surrounding nucleons. The efficiency of this thermal matching depends on the involved heat capacities, which can be expressed in terms of D for the electronic dipolar, and ω_I for the nuclear Zeeman reservoir, where ω_I is the nuclear Larmor frequency with respect to the external field. The thermal matching is optimal, when the heat capacities, hence the corresponding transition energies, are of the same magnitude, thus $g_e \mu_B D \approx \hbar \omega_I$.

2.1. Provotorov theory

The great success of the spin-temperature theory in describing the DNP mechanism is, that it allows to illustrate the DNP like above, but also to treat the different steps of cooling (heating) and thermal matching more quantitatively. The most important relations concerning the cooling (heating) of the electronic dipolar reservoir, the coupling to the nuclear Zeeman reservoir and, consequently, the maximum reachable nuclear polarization, are introduced briefly in the following.

First, one has to define the so-called inverse spin temperatures, namely, the inverse Zeeman temperature α, and the inverse spin-spin or dipolar temperature β:

$$\alpha = \frac{\hbar}{kT_{Ze}} \quad , \quad \beta = \frac{\hbar}{kT_D} \quad . \tag{1}$$

Although the Provotorov theory in principle does not differentiate between electronic and nuclear spin systems, the inverse spin temperatures defined here refer to the electrons, in order to be in line with the illustrations given above. The Provotorov equations then describe the time expansion of the inverse electronic spin temperatures:

$$\frac{d\alpha}{dt} = -W(\alpha - \beta) - \frac{1}{t_Z}(\alpha - \alpha_L) \tag{2}$$

$$\frac{d\beta}{dt} = W \frac{\Delta^2}{D^2}(\alpha - \beta) - \frac{1}{t_D}(\beta - \beta_L) \tag{3}$$

Here, W denotes the Zeeman transition probability, and $\Delta = \omega_e - \omega$ is the difference between Larmor and microwave frequency. The second terms are phenomenological expressions, which describe the relaxation to the lattice, respectively. They depend on the Zeeman and dipolar relaxation times, t_Z and t_D, and on the difference between the inverse spin temperatures and the inverse lattice temperatures $\alpha_L = (\omega_e/\Delta)\beta_L$ and $\beta_L = \hbar/kT_L$.

¿From the stationary solutions ($d\alpha/dt = d\beta/dt = 0$) of the Provotorov equations, one obtains an expression for the maximum inverse spin-spin temperature,

$$|\beta_{max}| = \frac{\omega_e}{2D}\sqrt{\frac{t_D}{t_Z}}\beta_L \quad , \qquad (4)$$

by assuming a fully saturating microwave field. Hence, the smaller the dipolar width D (i.e., the heat capacity) and the weaker the coupling to the lattice (i.e., the longer t_D), the better is the cooling (heating) efficiency of the dipolar reservoir.

In the next step, the coupling of the nuclear Zeeman reservoir to the electronic dipolar reservoir has to be described. This can be done with respect to the sum of both heat capacities and the contributing relaxation processes. By assuming that the nuclear Zeeman reservoir is completely isolated from the lattice, so that it can only relax via the dipolar reservoir, the increase of the heat capacity (compared to that of the dipolar reservoir alone) is equalized by the increase of the effective relaxation time so that, consequently, relation (4) holds for the combined dipolar and nuclear Zeeman reservoir. In fact, the nuclei are not completely isolated, but their coupling to the lattice is significantly weaker than that of the more versatile electrons. By summing up all relaxation processes of the nuclear Zeeman reservoir, which do not proceed via the dipolar reservoir, in a leakage factor f, relation (4) reads

$$|\beta_{max}| = \frac{\omega_e}{2D\sqrt{1+f}}\sqrt{\frac{t_D}{t_Z}}\beta_L \quad . \qquad (5)$$

Thus, the maximum absolute value for the inverse spin temperature of the combined dipolar and nuclear Zeeman reservoir is reduced by a factor of $\sqrt{1+f}$ due to nuclear relaxation processes, mainly nuclear spin-lattice relaxation.

2.2. Nuclear polarization

The important result of the Provotorov theory is, that the maximum reachable polarization of nuclei with spin I is directly proportional to the maximum inverse spin temperature:

$$P_{I,max} = \frac{I+1}{3} \beta_{max}\, \omega_I \qquad (6)$$

Even though this is a very suggestive result, there is a certain difficulty in applying it to a real polarized target: the Provotorov equations have been derived from an expansion of the density matrix under the assumption $\mu B \ll kT$ (the so-called high temperature approximation). But, in order to realize acceptable polarization values, a polarized target has to be operated at very low temperatures, which means, in the sub-Kelvin temperature regime. Here, the Provotorov theory is not valid in full generality, meaning that the system of paramagnetic electrons has to be assumed to fulfill additional restrictive requirements in order to translate the result (6) to real polarized target conditions[6]. For example, one of these requirements has shown to be, that the electronic dipolar interaction is negligible compared to inhomogeneous interactions (like hyperfine interaction), but, on the other hand, has to be strong enough to generate a uniform spin-spin temperature.

Most of the common target materials fulfill these supplementary requirements – at least approximatively. For them, the maximum reachable nuclear polarization at low temperatures follows from (6) by replacing the factor $(I+1)/3$ by the corresponding Brillouin function ($\eta = t_Z/t_D$):

$$P_{I,max} = \mathcal{B}_I \left(I\beta_L\, \omega_e\, \frac{\omega_I}{2D}\, \frac{1}{\sqrt{\eta(1+f)}} \right) \qquad (7)$$

The meaning of this result is, that for an improvement of the nuclear polarization, besides the 'external' parameters temperature (β_L) and magnetic field (ω_e), the 'internal' parameters ω_I and D have to be kept in mind. As a result of the specialization towards low temperatures, the parameter D, which originally denoted the energy width of the electronic dipolar reservoir, now also contains contributions of inhomogeneous interactions (like hyperfine interaction) and therefore can be identified with the complete width of the ESR line. Hence, in terms of the spin-temperature theory, the ESR linewidth should be minimized until it matches the nuclear Zeeman energy.

3. Contributions to the ESR linewidth

The width of the ESR line is governed by the interactions of the paramagnetic electrons with each other, their surroundings, and the external magnetic field. Besides a small contribution of homogeneous line broadening mechanisms like dipolar spin-spin interaction, the linewidth of spin systems in solids is usually dominated by inhomogeneous effects which occur, if the spins are subject to the corresponding interaction nonuniformly, meaning systematically. In this case, the static external or the time averaged local magnetic field is different for different spins or groups of spins. Depending on the strength of the corresponding interaction, this may lead to the formation of so-called spin packets with different Larmor frequencies, resulting in a corresponding structure of the ESR line. Besides inhomogeneities of the external magnetic field, which should not play a noticeable role in well designed ESR spectrometers, the hyperfine interaction and the so-called g-anisotropy are the most important inhomogeneous broadening mechanisms.

3.1. Hyperfine interaction

The interaction between an electronic (S) and nuclear spin (I) leads to the hyperfine structure (HFS), the splitting into the $(2I + 1)$ magnetic states of the combined system. The paramagnetic electrons discussed here can interact with various adjacent nuclei – depending on the structure of the paramagnetic center. Independently of its special character (isotropic like in ^{6}LiD or anisotropic, which is the usual case), the hyperfine interaction can be described in terms of the hyperfine tensor **A**. Its Hamiltonian reads

$$\mathcal{H}_{hfs} = \vec{S} \cdot \mathbf{A} \cdot \vec{I} \qquad (8)$$

and is independent of the external magnetic field.

3.2. Anisotropy of the g-tensor

Inhomogeneous local fields like electric fields caused by the symmetry or by defects of the lattice can lead to a degenerated symmetry of the electronic wave function (quenched spin-orbit coupling or quenched angular momenta) and, consequently, to a distribution of the Larmor frequencies depending on the orientation of the crystal or molecular axes with respect to the external field. Thus, the g-factor becomes a tensor and the Hamiltonian for the interaction of the electron with the external field reads

$$\mathcal{H} = \mu_B \vec{B} \cdot \mathbf{g} \cdot \vec{S} \quad . \qquad (9)$$

In contrast to the HFS broadening, the magnitude of the broadening due to g-anisotropy is field dependent.

In most of the polarized target materials HFS as well as g-anisotropy contribute to the ESR linewidth, so that one has to deal with a superposition of field independent and field dependent broadening effects. Thus, an extrapolation of the results of linewidth measurements (e. g., yielded with a X-band spectrometer) on the magnetic field scale is only possible, if the individual contributions to the linewidth are known. The fact that this in general is not the case and, on the other hand, that the ESR linewidth is a crucial parameter with respect to the polarization behavior, emphasizes the importance of ESR measurements at DNP relevant magnetic fields.

4. Results and discussion

The V-band spectrometer, working at DNP conditions at $1\,K$ and $2.5\,T$ is described in Refs. 4 and 8, as well as the ESR-results of various target materials obtained with this spectrometer.

The ESR linewidth of the deuterated target materials at $2.5\,T$ varies within one order of magnitude, from the extremely broad signal of EDBA on the one hand to the extremely narrow signal of the trityl radical OX 063 on the other hand. Except for the ^{6}LiD (HFS only) and the nitroxide radicals (HFS and g-anisotropy), the lineshape and, consequently, the linewidth of all other materials, is governed by g-anisotropy at these high magnetic fields. The results for the linewidth and the relative g-anisotropy are listed in Table 1, and, additionally, the highest measured deuteron polarization (at $2.5\,T$).

The polarization values for d-butanol (EDBA, TEMPO, Porphyrexide) and irradiated d-butanol are taken from Ref. 9. It has to be remarked that in EDBA doped materials the deuteron polarization can be rised up to $40-50\,\%$ using microwave frequency modulation[10], but for better comparability the value reached without modulation is given here. The value for ND_3 is taken from Ref. 11, for ^{6}LiD from Ref. 12, and for d-butanol (Finland D36) as well as for d-propanediol (OX 063) from Ref. 3.

These values of the maximum deuteron polarization show a very clean correlation with the ESR linewidth: the narrower the ESR line is, the higher is the maximum deuteron polarization. This correlation is in agreement with the conclusion of the spin-temperature theory, that for high deuteron polarizations the ESR linewidth has to be minimized. The corresponding plot is shown in Fig. 2. From this it can be seen, that there is not only a qualitative agreement between data and spin-temperature theory, but that the spin-temperature theory describes the data even in a semi-quantitative way: the dashed line represents the prediction (7) of the spin-temperature theory for deuterons with an assumed correction for the nuclear relaxation of $\sqrt{\eta\,(1+f)} = 3.5$. Although the correction is of the

Table 1. Relative g-anisotropy, ESR linewidth and maximum measured deuteron polarization of the investigated deuterated target materials at 2.5 T.

Material	Radical	$\Delta g/\bar{g}\,[10^{-3}]$	$FWHM\,[mT]$	$P_{D,max}$
D-Butanol	EDBA	5.98 ± 0.03	12.30 ± 0.20	26%
D-Butanol	TEMPO	3.61 ± 0.13	5.25 ± 0.15	34%
D-Butanol	Porphyrexide	4.01 ± 0.15	5.20 ± 0.23	32%
$^{14}ND_3$	$^{14}\dot{N}D_2$	$\approx 2...3$	4.80 ± 0.20	44%
$^{15}ND_3$	$^{15}\dot{N}D_2$	$\approx 2...3$	3.95 ± 0.15	-
D-Butanol	Hydroxyalkyl	1.25 ± 0.04	3.10 ± 0.20	55%
^{6}LiD	F-center	0.0	1.80 ± 0.01	57%
D-Butanol	Finland D36	0.50 ± 0.01	1.28 ± 0.03	79%
D-Propan.	Finland H36	0.47 ± 0.01	0.97 ± 0.04	-
D-Propan.	OX063	0.28 ± 0.01	0.86 ± 0.03	81%

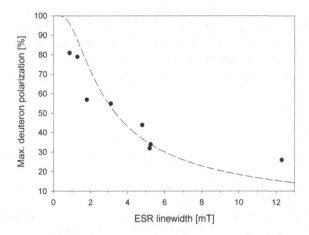

Figure 2. Maximum measured deuteron polarization as a function of the ESR linewidth at 2.5 T. Dashed line: spin-temperature theory prediction with $\sqrt{\eta(1+f)} = 3.5$.

right magnitude,[1,2] this specific material parameter is here assumed to be uniform for all materials, what obviously is a rough approximation. Since it seems to be impossible to exactly quantify the correction factor f for the different materials, it will not be possible to apply prediction (7) really quantitatively to a certain material. Nevertheless, the agreement between the data and this theoretical curve is impressive and points out the basic message of this comparison between polarization and ESR measurements on the deuterated target materials: due to their significantly small magnetic moment, the polarization of deuterons can be dramatically increased by using paramagnetic centers with minimal ESR linewidths (down to the theoretical optimum of about $0.6\,mT$) for the DNP.

5. Summary

These new developments in the field of deuterated polarized solid target materials will allow polarization experiments to be performed on the deuteron and neutron with a much higher precision than was previously possible. Doubling the maximum polarization means doubling the statistical accuracy for the same measuring time. Alternatively, for a given accuracy, the required measuring time is reduced by a factor of four. For these reasons, trityl-doped (Finland D36) d-butanol with its maximum d-polarization of 80 % was successfully used for the neutron part of an experiment to study the Gerasimov-Drell-Hearn sum rule at the Mainz microtron (MAMI), using a polarized tagged photon beam. For the first time, after more than 40 years of polarized solid targets, it is now possible to perform these kinds of experiments with low-intensity beams to the same precision and over the same time-span, no matter whether the proton or the neutron is the subject of investigation.

References

1. A. Abragam, *Principles of Nuclear Magnetism* (Oxford University Press, 1961)
2. M. Goldman, *Spin Temperature and Nuclear Magnetic Resonance in Solids* (Oxford University Press, 1970)
3. S. Goertz, J. Harmsen, J. Heckmann, Ch. Hess, W. Meyer, E. Radtke, and G. Reicherz, Nucl. Instr. and Methods A526 (2004) 43-52
4. J. Heckmann, PhD thesis, Bochum, Germany (2004)
5. M. Borghini, in *Proceedings of the 2nd International Conference on Polarized Targets, Berkeley, 1971*, edited by G. Shapiro (Berkeley, USA, 1971)
6. A. Abragam, M. Goldman, Rep. Prog. Phys. 41 (1978) 395
7. S. Goertz, Nucl. Instr. and Methods A526 (2004) 28-42
8. J. Heckmann, S. Goertz, W. Meyer, E. Radtke, and G. Reicherz, Nucl. Instr. and Methods A526 (2004) 110-116
9. J. Harmsen, PhD thesis, Bochum, Germany (2002)
10. Y. Kisselev, Nucl. Instr. and Methods A356 (1995) 99-101
11. W. Meyer, Habilitation thesis, Bonn, Germany (1988)
12. K. Kondo et al., Nucl. Instr. and Methods A526 (2004) 70-75

THE Sb, LiIO$_3$ AND HIO$_3$ ALIGNED NUCLEAR TARGETS FOR INVESTIGATION OF TIME REVERSAL INVARIANCE VIOLATION *

A. G. BEDA

Institute for Theoretical and Experimental Physics,
Moscow, Russia
E-mail: beda@itep.ru

L. D. IVANOVA

Institute for Metallurgy and Material Investigation RAS,
Moscow, Russia

The use of aligned nuclear targets for investigation of TRIV has a great discovery potential due to the large enhancement of TRIV effects in compound resonances of nuclei. The appropriate target materials of HIO$_3$, LiIO$_3$ and Sb single crystals in which the I and Sb nuclei can be aligned by brute force method at millikelvin temperatures were proposed in this work. The single crystals of required sizes were grown from HIO$_3$, LiIO$_3$ and metallic Sb and the construction of dilution refrigerator that is precooled by two stage pulse-tube refrigerator without any cryoliquids was developed. The use of proposed targets at the new neutron spallation source (JSNS,Japan) will make possible to discover TRIV or decrease the present limit on the intensity of parity conserving time violating interaction by two-three order of magnitude.

1. Introduction

The polarized neutron beams are very appropriate tool for investigation of fundamental symmetries violation (spatial parity and time invariance) in the neutron- nuclei interactions due to large enhancement of the effects of violation in the compound resonances of nuclei in comparison with its values in elementary nucleon-nucleon interactions. One can use also neutron-nuclei interaction for investigation of time invariance violation since the neutron forward scattering amplitude includes two T-non-invariant correlations : three-fold ($\mathbf{s}[\mathbf{p} \times \mathbf{I}]$) and five-fold ($\mathbf{s}[\mathbf{p} \times \mathbf{I}])(\mathbf{pI})$, here $\mathbf{s}$, $\mathbf{I}$ and $\mathbf{p}$ are

---
*This work is supported by RFBR grant N 03-02-16050.

the neutron spin, spin of the target nucleus and the neutron momentum accordingly. In the first case in the experiment it is necessary to use polarized and the in the second case aligned nuclear targets. Long ago it has been shown [1,2], that the values of T invariance violation effects can be enhanced in the p-wave compound resonances analogously to the enhancement of parity violation effects. In spite of the fact that the first proposals to use as three-fold [3,4] so the five-fold [5,6] correlations for investigation of T invariance violation have been made more than twenty years ago, real progress in this field was absent for lack of appropriate oriented nuclear targets. In this work the aligned nuclear targets are proposed where the Sb and I nuclei can be aligned by brute force method.

2. Aligned nuclei

An ensemble of nuclei with spin $I \geq 1$ is said to be aligned if the parameter of alignment

$$p_2(I) = \frac{3\langle m^2 \rangle - I(I-1)}{I(2I-1)} \qquad (1)$$

is not zero. Here m is the projection of spin I of a nucleus on the axis of alignment and $\langle m^2 \rangle = \sum_m m^2 n_m / \sum_m n_m$, n_m is the population of substate m. If the quadrupole nuclei is in the electric field gradient in the single crystal then the ground state of a nucleus with spin I is split due to an interaction of quadrupole nuclear moment with an electric field gradient (EFG) of the crystal. In this case the energies of substates determined by quadrupole interaction are defined as

$$E_m = \frac{eQq(3m^2 - I(I+1))}{4I(2I-1)}. \qquad (2)$$

Here eQq is a constant of quadrupole coupling. If the spins are in equilibrium at the temperature T the population distribution n_m over substates is given by the Boltzman law. The typical equilibrium value of the alignment $p_2(I)$ for heavy nuclei at $T = 0.5$ K is about 0.5 %. It is necessary to emphasize that in this case there is no necessity to apply the magnetic field and the direction of the quantization axis coincides with the main axis of the crystal electric field. The higher degree of nuclear alignment can be achieved by dynamical nuclear alignment method [7] or in some cases by brute force method – by cooling of target material to within several tens of millikelvin.

The use of brute force method is quite perspective now since during last years there is an impressive progress in the field of low temperature technique. The new type of the dilution refrigerator was developed [8] which is precooled to a temperature 3-4 K by a commercial two-stage pulse-tube refrigerator (PTR) and thus no cryoliquids are needed to operate such millikelvin cooler. In addition, the refrigerator discussed above is economical and convenient since it can operate in automatic mode. It can also provide rather high refrigeration capacity which makes possible to cool down the targets of large mass. Besides, there is no loss of neutron beam time due to refilling of liquid helium.

3. The aligned nuclear targets involving $LiIO_3$, HIO_3 and Sb single crystals

The value of the time invariance violation effect p_T is proportional to

$$p_T = k \cdot l, \tag{3}$$

here factor k includes nuclear characteristics, l is the target thickness in cm. The statistical error Δ of the measurements is following

$$\Delta = \frac{1}{\sqrt{I_0 S T e^{-l/\lambda}}}. \tag{4}$$

Here I_0 is the neutron density (n/cm^2·s), S is target area in cm^2, T is measurement time and λ is a neutron free path. The necessary condition for observation of the effect is $p_T/\Delta \geq 3$.

The large number of p-resonances in the Sb^{121}, Sb^{123} and I^{127} nuclei makes it possible to perform the statistical estimation of the value of the p_T effects and deformation effects in the p-resonances of the above nuclei. This estimation has been carried out in [9]. It turned out that the p_T values in the p-resonances are two order of magnitude as great as the statistical error Δ obtained by TRIPLE collaboration at the measurement of P-odd effects in p-resonances at the spallation neutron source LANCE (Los Alamos, USA) [10]. Then we developed the method of the growing of large volume single crystals and have grown Sb single crystal with volume more than 100 cm^3. As to iodine, presently the methods of growing of single crystals of $LiIO_3$ and HIO_3 with the volume of several hundreds cm^3 were developed in the ifrared field. The value of quadrupole coupling constant in these single crystals is about 1000 MHz and the degree of alignment of I nuclei will be about 40 % at 20 mK.

Once the problem of crystal growing was solved, the pulse tube based dilution refrigerator for aligned targets was developed [11]. It provides possibility to cool down the large volume targets to several tens of millikelvin.

In this case the single crystals must be cut into plates with thickness not more 10 mm to provide effective cooling by ^{3}He-^{4}He mixture circulating between the plates.

The calculation of targets heating on polarized neutron beam irradiation were performed for LiIO$_3$, HIO$_3$ and Sb targets . The volume of targets is taken equal to 125 cm^3 [11] and the distance to the center of neutron spallation source JSNS is taken to 10 m. The results of calculation is shown in Table 1.

Table 1. Target heating.

Target	Reaction	Heating power, μW
He^3	(n,p)	27
LiIO$_3$		
^{6}Li	(n,α)	100
^{127}I	(n,γ)	14
^{127}I	decay	9
	Total	**123**
HIO$_3$		
^{1}H	(n,n')	35
^{1}H	(n,γ)	4
^{127}I	(n,γ)	17
^{127}I	decay	11
	Total	**67**
Sb	(n,γ)	68
^{121}Sb	decay	18
^{123}Sb	decay	15
	Total	**101**

The dependence of the degree of the alignment of I and Sb nuclei on temperature is shown in Fig. 1. In the same figure it is shown the dependence of cooling power Q on temperature for the pulse-tube based refrigerator using PTR 410 (Q = 1 W at 4.2 K, Cryomech, USA) with continuous throughput of 800 μ mol/s. By means of Table 1 and Fig. 1 one can estimate that under neutron irradiation of aligned target of the HIO$_3$ single crystal with volume of 30 cm^3 this target can be cool down to temperature about 40 mK. At that the degree of alignment of I nuclei will be about 20 %.

According to estimation one can discover the time invariance violation effect or to lower the present bound on the intensity of P-conserving T-violating interaction by two-three order of magnitude in the experiment at JSNS with the use of the aligned targets of LiIO$_3$, HIO$_3$ and Sb with volume about 30 cm^3.

4. Conclusion

So, presently there is a possibility to realize the experiment on the investigation of time invariance violation in neutron-nuclei interaction with the use of the aligned nuclear targets. The appropriate target materials of LiIO$_3$, HIO$_3$ and Sb single crystals are proposed in which I and Sb nuclei can be aligned by brute force method. The construction of present day pulse tube based dilution refrigerator for nuclear alignment is developed.

References

1. V. E. Bunakov, V. P. Gudkov, Nucl. Phys. A 401 (1983) 93.
2. V. E. Bunakov, Phys. Rev. Let., 60 (1988) 2250.
3. P. K. Kabir, Phys. Rev. D 25 (1982) 2013.
4. L. Stodolsky, Nucl. Phys. B 197 (1982) 213.
5. V. G. Baryshevcky, Sov. J. Nucl. Phys. 38 (1983) 1162.
6. A. L. Barabanov, Sov. J. Nucl. Phys. 44 (1986) 755.
7. V. A. Atsarkin, A. L. Barabanov, A. G. Beda, V. V. Novitsky, Nucl. Instr. Meth. A, 440 (2000) 626.
8. K. Uhlig, Cryogenics 42 (2002) 73
9. A. L. Barabanov, A. G. Beda, J. Phys. G.: Nucl. Part. Phys. 331 (2005) 161
10. G. E. Mitchell et al., Phys.Rep. 354 (2001) 157.
11. A. G. Beda, A. N. Chernikov, Czechoslovak Journal of Physics 55 (2005) Suppl. B, 357.

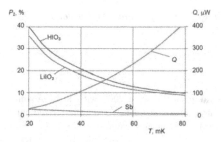

Figure 1. Dependence of the refrigerator cooling power (Q) and nuclear alignment (p_2) for I nucleus (in single crystals of HIO$_3$ and LiIO$_3$ and for Sb nucleus (in Sb single crystal) on temperature T.

FUTURE ACTIVITIES OF THE COMPASS POLARIZED TARGET

N. DOSHITA*, J. HECKMANN, CH. HESS, Y. KISSELEV, J. KOIVUNIEMI,
K. KONDO, W. MEYER AND G. REICHERZ
Physics Department, University of Bochum, 44780 Bochum, Germany

J. BALL, A. MAGNON, C. MARCHAND AND J.M. LE GOFF
CEA Saclay, DAPNIA, 91191 Gif-sur-Yvette, France

G. BAUM AND F. GAUTHERON
Physics Department, University of Bielefeld, 33501 Bielefeld, Germany

T. HASEGAWA AND T. MATSUDA
Faclty of Engineering, Miyazaki University,889-2192 Miyazaki, Japan

N. HORIKAWA
*Research Institute for Science and Technology, Chubu University,
487-8501 Kasugai, Japan*

T. IWATA
*Department of Physics, Faculty of Science, Yamagata University,
990-8560 Yamagata, Japan*

The COMPASS experiment has been taking data since 2002 with a large solid polarized target and a high intensity muon beam at CERN[1]. The polarized target apparatus is upgraded in 2006, specially a large acceptance 2.5 T solenoid superconductive magnet is installed.

---

*Email: norihiro.doshita@cern.ch

1. Introduction

The COMPASS experiment with the polarized target and the polarized muon beam of 160 GeV at CERN began in 2002 in order to determine the contribution of the gluon polarization in the polarization of the nucleon by measuring the double spin cross section asymmetries for reactions via the Photon-Gluon Fusion (PGF) process in two different processes. The first one is the hadron pairs production with large transverse momentum, that has a higher statistical accuracy but is affected by larger systematic uncertainties due to background processes. The second one is the open charm production which can be identified as the cleanest and most direct tool to access the gluon polarization because there is little intrinsic charm in the nucleon and the open charm production are unlikely to be produced by processes other than the PGF processes. The charm production can be recognized by detecting charmed particles in the final state (experimentally D^0 and $\bar{D}^0$). However, the statistics of the charm production is much less than hadron pairs production.

The experimentally measurable double spin asymmetry A_{raw} is

$$A_{\text{raw}} = \frac{N^{\leftrightarrows} - N^{\leftleftarrows}}{N^{\leftrightarrows} + N^{\leftleftarrows}} = P_B P_T f A_{\|} \qquad (1)$$

where $N^{\leftrightarrows}$ ($N^{\leftleftarrows}$) is the counting rate of the hadron pair production or the open charm production in the anti-parallel (parallel) spin configuration between the beam and the target, P_B is the beam polarization, P_T is the target polarization and f is the dilution factor. The gluon polarization can be extracted from $A_{\|}$[2,3].

2. The new polarized target apparatus

The target apparatus from 2002 to 2004 consisted of a 2.5 T superconductive solenoid magnet which has a 69 mrad acceptance, two 60 cm long targets of ^{6}LiD inside a 160 cm long mixing chamber of a dilution refrigerator, two microwave systems and a 10 NMR signals detection system[4]. The COMPASS is upgrading the polarized target apparatus for the run from 2006 onward, as following,

(1) *new 180 mrad acceptance superconducting magnet*
(2) *3-target-cells set up*
(3) *large diameter microwave cavity*

The final set up of the apparatus is shown in Fig. 1. Compared with the former magnet, the new magnet with 180 mrad acceptance makes the

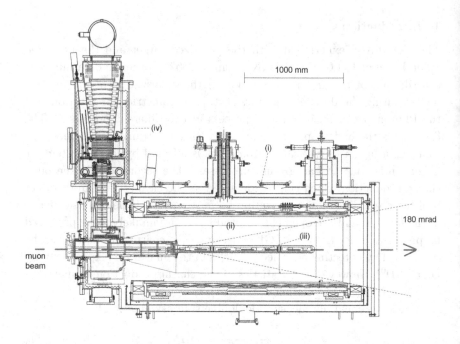

Figure 1. Side view of the new COMPASS polarized target apparatus. (i) 180 mrad superconductive magnet, (ii) large diameter cavity, (iii) 3-target-cells, (iv) dilution refrigerator.

statistics of the charm production gain 30 % or more and improve the data quality of the hadron pair production.

A homogeneity of less than ±50 ppm over the long target region is required in order to obtain high polarizations which needs a precise microwave frequency corresponding to the magnetic field. The result of the homogeneity test at 2.5 T with trim coils in September 2005 at CEA Saclay was less than ±30 ppm over the target region, as shown in Fig. 2.

The magnet was mounted on the target platform of the COMPASS experiment in December 2005 and the cool down was started in the beginning of January 2006.

The 3-target-cells set up, shown in Fig. 3, instead of 2-target-cells makes two pairs of 2-target-cells, that helps to reduce false asymmetry. A new microwave cavity[5] is needed for the new system fitting to the new magnet.

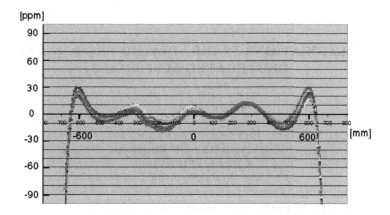

Figure 2. Profile of the 180 mrad acceptance solenoid magnet homogeneity. The target region is from -65 cm to 65 cm.

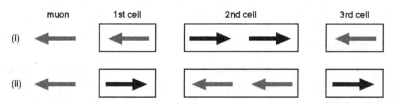

Figure 3. Two cases of spin configurations with the muon beam and the 3-target-cells set up. The arrows indicate the spin orientations. The most upstream cell (1st cell) is 30 cm long, middle one (2nd cell) 60 cm and downstream (3rd cell) 30 cm. One can realize two pairs of the 2-target-cells, one pair is the 1st cell and a half of the upstream side of the 2nd cell and the other is the other half of the 2nd cell and the 3rd cell.

References

1. G.K. Mallot, *Nucl. Instr. and Meth.* **A518**, 121 (2004)
2. COMPASS Collaboration., *Phys. Lett.* **B633**, 25 (2006).
3. J. Pretz, *to be published in the proceedings of HEP2005 International Europhysics Conference on High Energy Physics*.
4. J. Ball et al., *Nucl. Instr. and Meth.* **A498**, 101 (2003).
5. Y. Kisselev et al., *in these proceedings*.

POLARIZATIONS IN IRRADIATED PROTON AND DEUTERATED MATERIALS

D. G. CRABB
PHYSICS DEPARTMENT
UNIVERSITY OF VIRGINIA
CHARLOTTESVILLE
VA 22903, USA

Polarizations from a series of irradiations of deuterated and hydrogenous materials are presented. Irradiated deuterated butanol has the best performance with polarizations of 50.5% and -47.6% at 5.0 T, rising to 62.4% and -49.6% at 6.55 T. Deuterated 1-pentanol has a similar performance at 5 T. The results from hydrogenated materials are not as promising, giving values considerably less than obtained with standard chemical doping.

1. Introduction

The University of Virginia Target Group has been conducting a systematic study [1,2] of the proton and deuteron polarizations obtained after irradiation of various polarized target materials. The polarization measurements were all made with a ^{4}He evaporation refrigerator operating at around 1K and at magnetic fields of 2.5 T, 5.0 T and 6.55 T. Subsequent measurements of the polarization as a function of beam dose were all made at 5.0 T. For the deuterated materials, d-butanol (C_4D_9OH) gave polarizations of better than 50% at 5 T and over 60% at 6.55 T, close to a factor of 2 better than obtained with chemically doped materials (e.g., with TEMPO). However the proton polarizations were significantly lower than obtained with chemical dopants.

All the irradiations were carried out at the MIRF facility at the NIST laboratory in Gaithersburg, MD [a], with an electron linac operating at an energy of 19 MeV and with beam currents in the range 4 - 15 μA. Because the irradiations are carried out with the sample immersed in liquid Argon, large variations in the dose deposited in the target material can occur with variations in the target position. We have used the EGS4 program to

---

[a]Medical and Industrial Radiation Facility

study this variation and to make a first step in trying to compare it with irradiations at other facilities. Figure 1. shows the energy deposited, as a function of distance into the irradiation dewar. The beam was tuned to irradiate the front face of the dewar with a uniform coverage. For this work we define the radiation dose in units of electrons cm^{-2} incident on the dewar face.

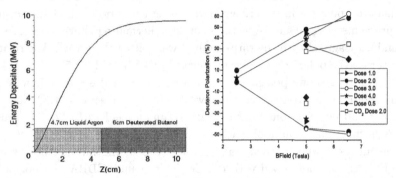

Figure 1. Energy deposited as a function of distance through the dewar by a 19 MeV electron beam incident on the face of the dewar.

Figure 2. Polarizations of d-butanol at different doses and CD_2 at one dose, at magnetic fields of 2.5, 5.0 and 6.55 T.

2. Material Preparation and Irradiation

The alcohols were frozen into 1 - 2 mm beads by a controlled drip from a burette into a liquid nitrogen bath and then collected and stored. The borane ammonia (BH_3NH_3), which is a powder, was pressed into thin discs, while the deuterated polyethylene (CD_2) was melted into an ingot and then cut into 1 - 2 mm pieces.

The material was put into an aluminum wire basket, mounted on the end of an aluminum tube which had a centering disc attached so as to ensure that the basket was positioned in the dewar on the beam line. An adjustment for height was also possible. The stick was placed in the liquid Argon in the dewar and the irradiation begun. For long targets the basket was rotated several times to ensure a more uniform irradiation of the material. For the deuterated material in these studies, only a small sample for a given condition was used.

3. Polarization Results

3.1. *Deuterons*

Figure 2 shows the results first obtained with deuterated butanol and CD_2[1], while Fig. 3 is the compilation of polarization results, at 5 T, obtained from all the deuterated materials irradiated for this study. The polarizations were reached after enhancing between one and a half and two hours. The deuteron polarizations were evaluated using the ratio method [3]: two measurements were made using the standard thermal equilibrium calibration and they agreed with the ratio method well within the errors. The relative error on the polarization measurements ranged from 3% - 5% .

Figure 2 shows polarizations for a sampling of doses as a function of magnetic field. Apart from the lowest d-butanol dose (0.5 10^{15} electrons cm^{-2}) all polarizations increase with magnetic field with the 3 10^{15} electrons cm^{-2} dosed butanol reaching 62.5% and - 49.6% at 6.55 T. For CD_2, the highest value reached was 34.4 %. Unlike the butanol doped with TEMPO, butanol doped with the Cr V compound, EDBA, did not polarize at either 5.0 T or 6.55 T in agreement with earlier results [4]. This has now been explained [5] and will be addressed later.

Figure 3 shows the polarizations obtained at several doses for d-butanol, CD_2 and, 1-d-pentanol $(CD_3(CD_2)_4OH)$.' There is a broad maximum in polarization for the d-butanol, between 2 and 3 10^{15} electrons cm^{-2}, reaching a value of almost 50%. CD_2 improves slightly between doses of 1 and 2 10^{15} electrons cm^{-2} reaching a value of 27.6%, similar to that obtained with d-butanol doped with d-TEMPO. Generally the polarization value for the positive enhancement is larger than the negative enhancement in contradistinction to chemically doped butanol.

Included in Figure 3 are some further measurements on d-butanol: A sample was irradiated to 2.5 10^{15} electrons cm^{-2} and the previous samples at 2 and 3 10^{15} electrons cm^{-2} were repolarized for comparison. The 2.5 10^{15} sample was polarized to a deuteron polarization of +50.5% and -47.6% with a build up time of about two hours. The sample's polarization was measured with both the ratio and TE method: the result was that values were barely distinguishable and can hardly be seen in the figure. This sample was also polarized at 6.55T achieving a value of greater than 60% in agreement with our previous measurements. The polarizations of the 2 and 3 10^{15} samples are also shown with build up times of six and three hours, respectively. About 97% of the ultimate polarization was obtained in approximately three hours.

Finally, Figure 3 shows very recent data from deuterated 1-pentanol. For experimental reasons the irradiation basket had to be changed and was a different length: EGS4 was used to estimate the effective dose, normalized to the incident beam flux. A butanol sample was also polarized to check on the normalization. The d-pentanol has a performance, within the normalization error, similar to that of d-butanol. This requires more study but d-pentanol appears to be a competing target candidate along with butanol.

3.2. Protons

Figure 4 shows the data accumulated for hydrogenous materials, namely butanol, polyethylene (CH_2) and BH_3NH_3. None of the polarizations approach those that can be obtained routinely with chemically doped alcohols (∼ 80%), or with irradiated ammonia.

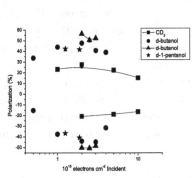

Figure 3. Deuteron polarizations for d-butanol, CD_2 and 1-d-pentanol as a function of beam dose.

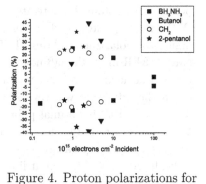

Figure 4. Proton polarizations for butanol, polyethylene (CH_2), borane ammonia (BH_3NH_3) and 2-pentanol as a function of beam dose.

Both BH_3NH_3 and CH_2 have broad maxima around 25% and -23% polarization. In contrast butanol and 2-pentanol have relatively narrow peaks with the highest values reaching around 40%. It is not clear from these measurements where the peak values are but it is unlikely that they will approach those seen in chemically (TEMPO) doped butanol and pentanol. Material made by our U. of Michigan colleagues gave values of 72% and -81% for butanol, measured in a similar system to ours, and 62% and -67%

for 3-pentanol [6]. These values were later confirmed in our apparatus.

4. Conclusion and Future Measurements

The results described above show for the conditions of 5 T and 1 K, the proton polarizations obtained with irradiated materials other than ammonia, do not approach those obtained through chemical doping. For example, butanol doped with TEMPO gave polarizations of greater than 80%, whereas the best seen with irradiated butanol was about 45%.

On the other hand, the irradiated deuterated materials surpass that of the chemically doped materials by approximately a factor of two, at least for butanol and pentanol, while CD_2 is about the same. Werner Meyer, in a presentation at this workshop, showed that from a series of EPR studies at 2.5 T at Bochum University, it was found that the value of deuteron polarization obtained was related to the EPR line width of the radical used. Without going into any detail it was found that, the narrower the line, the higher the polarization obtained. The radicals from irradiation tend to have narrow EPR lines, though not the narrowest., while the EDBA radical has the widest, leading to very small polarizations at 5 T and 1 K.

In the future we will repolarize some of these materials at 2.5 T and 0.5 K to find their optimal polarizations under these different conditions. We will also irradiate some selected materials under liquid helium, either because of the temperature of their melting points or to work with different radicals than are obtained at liquid Argon temperatures.

References

1. D. G. Crabb, Nucl. Instr. and Meth. in Phys. Res., **A526**, 56 (2004).
2. D. G. Crabb, Proc.Third Int. Symp.on the GDH Sum Rule and its Extensions, Norfolk, 2004, Eds. S. Kuhn and J. P. Chen, World Scientific, pp 183 -189 (2005).
3. C. Dulya et al., Nucl. Instr. and Meth. **A398**, 109 (1997).
4. S. Trentalange et al., Proc. Symp. on High Energy Spin Physics, Bonn, 1990, eds., W. Meyer, E steffens and W. Thiel, Springer Verlag 1991, Vol. 2, (1991).
5. W. Meyer, these proceedings
6. Alyssa Delke, private communication

MAGNET AND BEAM STUDIES FOR THE JLAB HALL-B FROZEN SPIN POLARIZED TARGET*

O. DZYUBAK, C. DJALALI, S. STRAUCH AND D. TEDESCHI

Department of Physics and Astronomy
University of South Carolina
Columbia SC 29208, USA
E-mail: dzyubak@physics.sc.edu

Several experiments at Thomas Jefferson National Accelerator Facility (JLab) Hall-B will use a "Frozen Spin" polarized target. This mode requires two separate working regimes, the polarizing of the target and the holding of the polarization. Two crucial parameters that eventually impact the experimental setup are the homogeneity of the polarizing magnet and the extra heat load from the photon beam. We present the results of the magnet field homogeneity and beam heat deposition studies.

1. Introduction

Recently at JLab Hall-B five experiments using the CLAS detector[1] and a polarized target have been approved[2]. After taking into account all the requirements of these projects, a reasonable choice for a polarized target is to use the "Frozen Spin" mode. In this mode, the target material will be dynamically polarized outside the CLAS detector at $B = 5.0$ Tesla and $T = 1.0$ K. The field homogeneity of the polarizing magnet ultimately defines the enhanced polarization values. After the maximal polarization is achieved, the cryostat is turned to the "holding" mode and moved inside the CLAS detector. The holding field and the temperature define the relaxation time of the polarization. At fixed holding field $B = 0.5$ Tesla and $T = 50.0$ mK, the expected relaxation time is about $t = 200\text{-}300$ hours[3,4,5,6,7,8]. However an external heat load can increase the temperature causing an unacceptable reduction in the relaxation time resulting in frequent repolarization cycles.

The design work on the 50.0 mK dilution refrigerator[9] and the holding magnets[10,11] has been reported previously. The purpose of this work is to

---
*This work is supported by DOE/EPSCoR Grant # DE-FG02-02ER45959.

study the field homogeneity of the polarizing magnet and the photon beam heat deposition in the target material.

2. Field homogeneity of the polarizing magnet

The 5.0 Tesla magnet was built by Cryomagnetics, Inc. and will be used to dynamically polarize (DNP technique) the target material embedded with paramagnetic impurities (alcohols with Cr-V, irradiated ammonia, irradiated lithium, etc.). The DNP processes are defined by the behavior and the concentration of these paramagnetic centers (EPR-line shape and width σ_{EPR}). The experimental studies of the dependency "maximal polarization value vs. magnetic field" are summarized in Table 1. As can be seen from

Table 1. Target material studies.

Material	σ_{EPR}, MHz(Oe)	Magnet, Tesla	ΔH, Oe	N_{PC} $\times 10^{20} cm^3$	Ref.
Propanediol(EHBA-CrV)	750 (270)	5.0	270	1.2–2.5	[12]
Irradiated ammonia	200 (70)	2.7	90	1.0	[13,14,15,16]
Propanediol, Butanol(CrV)	230 (82)	2.7	82	0.1–0.5	[17]

the table and references cited therein, for commonly used target materials, the distance ΔH in units of a magnetic field between points where polarization reaches its maximal value ranges from 82.0 Oe for low concentrations of paramagnetic centers N_{PC} (meaning longer relaxation time) to 270.0 Oe for higher concentrations and to prevent any polarization leak during DNP, the field should be kept under resonance conditions within about 5.0 Oe. Our run conditions require the target to be a cylindrical cell 1.5 cm in diameter and 5.0 cm in length. To get the maximum values during the polarization process, the magnet field homogeneity should be better than 100 ppm over the target cell volume if the 5.0 Tesla polarizing field is used. For our measurements we used a Hall-Gaussmeter Model-450 from Lake Shore Cryotronics, Inc.[18] and a Digital NMR Teslameter PT-2025 from GMW Associates[19]. The NMR-probe was used without field gradient compensation coils; over regions with a high gradient, the Hall-probe was used instead[20]. The main results are presented in Figure 1. The measured homogeneity of the magnetic field over the target area is better than 40 ppm. Our tests showed that the polarizing magnet is very reliable and is capable to provide the optimal polarizing conditions for a variety of target materials. With this magnet we expect the average polarization value to

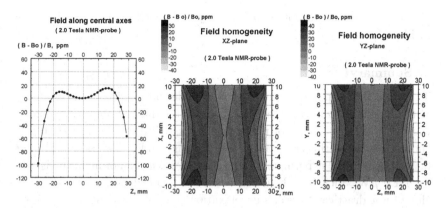

Figure 1. Polarizing magnet field homogeneity.

be 85-90%.

3. Beam heat deposition

Since the new target will be used only in photo-nuclear experiments, in addition to ammonia NH_3, butanol C_4H_9OH and lithium hydride LiH can also be used as target materials. The heat deposition in the target due

Table 2. Beam heat deposition.

Material	Density(g/cm^3)	Packing factor	Beam heat, μW
$C_4H_9OH + He$	0.94	0.62	0.51
$NH_3 + He$	0.85	0.58	0.40
$LiH + He$	0.82	0.55	0.14

to the interaction with the photon beam was simulated with the GEANT 3.21 code. The 1.0 GeV photon beam consisted of 10^7 γ/s. The target material was considered as spherical beads with about 2.0 mm in diameter immersed in liquid ^4He and neglecting ^3He component[21]. The packing factor values along with densities of materials at 77 K were taken from references[22,23]. Some parameters of the target materials together with the results of our simulations are presented in Table 2. It can be seen that for all target material candidates, the additional external heat load which comes from the photon beam does not exceed 1.0 μW. According to preliminary calculations, the cryostat cooling power in the holding mode (0.5 Tesla and 50.0 mK) is expected to be about 10.0 μW. Thus the relaxation time of polarization should not be affected by the heat load generated by the incident beam.

4. Summary

The JLab Hall-B polarizing magnet is capable to provide the homogeneity of the magnetic field over the target area better than 40 ppm which allows the use of a large variety of target materials. With an expected 10.0 μW cooling power of the cryostat in the holding mode, the additional heat deposition during experimental run coming from a photon beam is less then 1.0 μW. Such heat load can easily be handled by the cryostat and should not drastically affect the relaxation time of polarization. Work on the experimental setup for calibration and monitoring the polarization has been started in collaboration with the Kharkov Institute of Physics and Technology (Kharkov, Ukraine)[24,25].

References

1. B. Mecking et al., *Nucl. Instr. Meth. in Phys. Res.* **A503**, 513, (2003).
2. clasweb.jlab.org/frost/CLAS_FST_Experiments.html
3. N. S. Borisov et al., *Nucl. Instr. Meth. in Phys. Res.* **A345**, 421, (1994).
4. H. Dutz et al., *Nucl. Instr. Meth. in Phys. Res.* **A356**, 111, (1995).
5. N. A. Bazhanov et al., *Nucl. Instr. Meth. in Phys. Res.* **A372**, 349, (1996).
6. N. A. Bazhanov et al., *Nucl. Instr. Meth. in Phys. Res.* **A402**, 484, (1998).
7. R. Gehring et al., *Nucl. Instr. Meth. in Phys. Res.* **A418**, 233, (1998).
8. Ch. Bratke et al., *Nucl. Instr. Meth. in Phys. Res.* **A436**, 430, (1999).
9. C.D. Keith, M.L. Seely, O. Dzyubak, *GDH-2004: Proc. 3rd Intern. Symp. on GDH Sum Rule and its Extensions*, (World Scientific Pub., 2005), p.201.
10. O. Dzyubak, *CLAS-NOTE 2003-002*, (2003).
11. O. Dzyubak et al., *Nucl. Instr. Meth. in Phys. Res.* **A526**, 132, (2004).
12. A.A. Belyaev et al., *J. Applied Spectroscopy* **68(4)**, 623 (2001).
13. O. Dzyubak, *PhD Thesis*, (1989).
14. A.V. Vertij et al., *Sov. Phys. Dokl.* **35**, 899 (1990).
15. A.A. Belyaev et al., *AIP Conf. Proc.* **187**, 1336 (1989).
16. V.P. Androsov et al., *AIP Conf. Proc.* **187**, 1346 (1989).
17. N.C. Borisov et al., *Sov. Phys. JETP*, **60(6)**, 1291 (1984).
18. www.lakeshore.com/mag/ga/gm450po.html
19. www.gmw.com/magnetic_measurements/MetroLab/PT-2025_Specs.html
20. O. Dzyubak, C. Djalali, D. Tedeschi, *CLAS-NOTE 2004-023*, (2004).
21. O. Dzyubak, C. Djalali, D. Tedeschi, *CLAS-NOTE 2005-015*, (2005).
22. D. Adams et al., *Nucl. Instr. Meth. in Phys. Res.* **A437**, 23, (1999).
23. J. Ball et al., *Nucl. Instr. Meth. in Phys. Res.* **A498**, 101, (2003).
24. A. Byelyayev, O. Dzyubak, O. Lukhanin, *CLAS-NOTE 2006-002*, (2006).
25. A. Byelyayev, O. Dzyubak, O. Lukhanin, *CLAS-NOTE 2006-003*, (2006).

INVESTIGATIONS OF THE MULTIMODE CAVITY FOR THE COMPASS-MAGNET

Y. KISSELEV, N. DOSHITA, J. HECKMANN, J. KOIVUNIEMI, K. KONDO,
W. MEYER, G. REICHERZ

Physics Department, University of Bochum, 44780 Bochum, Germany.
E-mail: yuri.kisselev@cern.ch

G. BAUM, F. GAUTHERON
Physics Department, University of Bielefeld, 33501 Bielefeld, Germany.

J. BALL, A. MAGNON, S. PLATCHKOV
CEA Saclay, DAPNIA, 91191 Gif-sur-Yvette, France

G. MALLOT
CERN CH-1211, Geneve 23, Suisse/Switzerland

We study the effect of the frequency modulation (FM) of a microwave field on both the build-up time and the nuclear polarization in LiD target material. We show that FM-spectrum covers several modes in a large target cavity and therefore provides better spatial uniformity. The FM was successfully used in the COMPASS twin target where deuteron polarization of up to (+57% and -53%) was achieved.

DNP under the multimode microwave irradiation

The COMPASS target, made of LiD irradiated material, operates at 2.5 T solenoid field, with microwaves (MW) of $\lambda \cong 4$ mm wavelength and in the temperature range 0.06 to 0.25 K. The polarization of the target nuclei is obtained by using the Dynamic Nuclear Polarization (DNP) method the efficiency of which depends on the intensity and the spatial uniformity of the MW-field saturating the paramagnetic centers in the material.

We study the multimode excitation of a non-tunable microwave cavity in order to provide a high intensity and a good spatial uniformity of the MW field in the material as required for high nuclear polarizations. This is achieved by using a frequency modulation (FM) of the MW generator. The FM elementary wave $\sin\left[2\pi\left(\nu_0 + \Delta\nu \cdot \cos\left(2\pi f \cdot t\right)\right)t\right]$, where ν_0 denotes the carrier frequency, $\Delta\nu$ the frequency deviation and f the modulation frequency, has a symmetric amplitude spectrum relatively to the carrier with a bandwidth of $2 \cdot \Delta\nu$ and an interval between harmonics equal to f Ref. [1]. In our case $\nu_0 \approx 70$ GHz, $\Delta\nu \approx 5$ MHz, $f \approx 500$ Hz and the bandwidth of $2 \cdot \Delta\nu \approx 10$ MHz which is the most-used for target materials.

The total amount of TM and TE modes of an empty cylindrical cavity can be estimated by the formula of Ref. [2] for their frequencies f_{TM} and f_{TE} correspondingly

$$f_{TM} = \frac{3 \cdot 10^{10}}{2\pi}\sqrt{\left(\frac{p\pi}{l}\right)^2 + \left(\frac{\eta_{nm}}{R_0}\right)^2}, \; f_{TE} = \frac{3 \cdot 10^{10}}{2\pi}\sqrt{\left(\frac{p\pi}{l}\right)^2 + \left(\frac{\mu_{nm}}{R_0}\right)^2}, \tag{1}$$

where $p = 1, 2, 3 \cdots$, η_{nm} and μ_{nm} are respectively the roots of the Bessel functions and its derivatives, $l=100$ cm is the length and $R_0 = 11.5$ cm is the radius of our cavity. The highest modes were rejected as soon as the

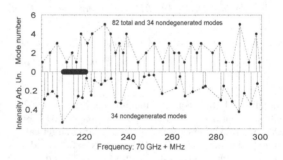

Figure 1. TM and TE modes (top) calculated from Eqs. (1). The solid line shows the bandwidth of FM-spectrum. Data (bottom) obtained with our receiver.

interval between adjacent modes exceeds the 70250 ± 50 MHz operating range. Under this condition the density of nondegenerated modes of ≈ 4 modes per 10 MHz was found in good agreement with the data. The bandwidth of the MW generator, the data and the modes calculated from Eq. (1) are plotted Figure 1. The cavity spectrum is investigated at room temperature using a microwave receiver. The receiver consists of a hybrid rigid cable-waveguide pickup antenna, a ferrite input modulator driven by a low frequency generator, a microwave detector and a display with a linear performance within the input power from 0 to 50 μW.

The multimode excitation must have an optimal matching between the feeding waveguide and the cavity within the broad bandwidth. Figure 2 (top) shows the coupling design in which the three-rectangular holes in the

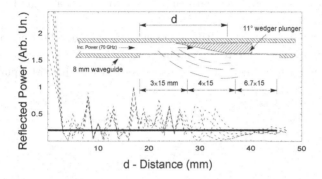

Figure 2. The design and dimensions of the coupler which enables the matching needed for DNP (top). Reflected power (bottom) over the hole length (d) for the five frequencies within 70250.0 ± 50 MHz. Solid line shows the the reflection from matching load.

broad side of Ka-waveguide are terminated by a wedge-shaped shorting plunger. At the matching opening ($d \geq 35$ mm), all the input power comes out from the waveguide and the reflection reaches its minimum values.

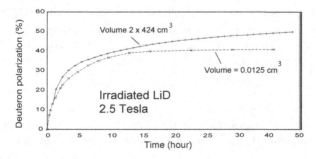

Figure 3. The DNP in the LiD material. The COMPASS data (2004) obtained with multimode excitation (top) are compared with the data for a single crystal Ref. [3].

Our studies show that FM has a negligible influence in the LiD up to ≈ 35% polarization. A polarization of about 50% can be reached also without FM in a small volume of a correctly irradiated LiD as shown in Ref. [4, 5], but shorter build-up time and ultimate polarizations of +57% and -53% in a large volume can only be reached if the multimode excitation is turned on. These features are demonstrated in Figure 3 where the deuteron build-up time and polarization in the LiD COMPASS target of 2×424 cm^3 under FM radiation are compared with the data obtained for a single LiD crystal of 0.0125 cm^3 (see Ref. [3]).

We suggest that the multimode operation enlarges the fraction of the nonlinear (with respect to the input power) resonant magnetic losses in comparison with linear dielectric losses. As a result the FM equalizes the

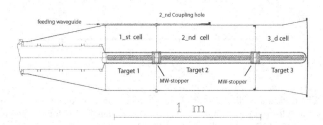

Figure 4. Planned three-cell target within MW-cavity of 40 cm diameter. The targets of (300 + 600 + 300) mm length are separated by the two microwave stoppers of 5 cm length each. Only one of three feeding waveguides is shown.

saturation of electron spins and it lowers the dielectric overheating of a material allowing shorter build-up time and higher ultimate polarizations.

Figure 4 presents the three cell target design for the COMPASS experiment. The use of three cells with the proposed geometry, instead of two cells, allows an order of magnitude reduction in systematic errors (See Refs. [6, 7] for more details).

References

1. J. Fagot et Ph. Magne, *La modulation de Fréquence Théorie et Application aux Faisceaux hertziens* , Société Française de documentation électronique, (1961).
2. David M. Pozar, *Microwave Engineering* , John Wiley and Sons. INC, 2nd Ed., 320 (1998).
3. V. Bouffard, Y. Roinel, P. Roubeau and A. Abragam, *J. Physique* **41**, 1447 (1980).
4. J. Ball et al., *Nucl. Instr. and Meth.* **A 498**, 101 (2003).
5. W. Meyer, *This Proceedings*.
6. Jean-Marc Le Goff, *COMPASS note*, **2004-3 CERN**, (2004).
7. J. Pretz, *COMPASS note*, **2004-11 CERN**, (2004).

DISTILLATION AND POLARIZATION OF HD*

S. BOUCHIGNY, J-P. DIDELEZ, G. ROUILLE

IN2P3, Institut de Physique Nucléaire, Orsay, France
E-mail: bouchign@ipno.in2p3.fr

The polarization (static or dynamic) of HD material, requires very pure HD samples, with H_2 and D_2 impurities concentrations smaller than 0.1%. At IPN Orsay, we have developed a new distillation apparatus, equipped with a mass spectrometer, allowing to reach such a level of purity and to measure the H_2 and D_2 contaminations down to 0.05%. The apparatus is described as well as the production of two HD samples, respectively suitable for static and dynamic polarization processes.

1. Introduction

Polarized HD targets are made of nearly pure HD molecules and have an outstanding "Dilution Factor", since all nuclear species in the target are polarizable [1]. Therefore, the polarization of HD molecules has been attempted by both the static and the dynamic methods. In 1967, A. Honig proposed a "Brute Force" (BF) method [2] by which the polarization of H and D nuclei obtained at very low temperature and high field (10mK and 15T), could be preserved at moderate temperatures and low fields for experimental use, so that recently, successful nuclear physics experiments with polarized HD targets could be performed [3]. In 1973, J.C. Solem investigated the Dynamic Nuclear Polarization (DNP) in radiation-dammaged solid HD, reaching 3.7% proton polarization and 0.3-0.4% for the vector polarization of deuterons [4]. More recent attempts could not improve the results obtained by Solem [5]. A detailed description of the BF and DNP methods can be found in Ref. [1]. In both cases, it is important to control the purity of the HD samples. For the BF method, the ideal HD sample should have some H_2 with a concentration of the order of 5×10^{-4} and typically $1 - 5 \times 10^{-4}$ of D_2. In the DNP process, the polarization of

---

*research supported in part by the european community through the i3hp network for the jra8 activity

free electrons created in the irradiated sample is transferred to the nuclear species by RF. To reach the high polarization of the nuclear species, their relaxation times T_1^H and T_1^D must be long compared to the electron one [6]. Therefore an HD sample suitable for DNP should contain as little as possible of H_2 and D_2, at the limit of the distillation performances. In this paper it will be shown that the required purity levels for HD samples can be reached with the distillation apparatus developed at Orsay and the corresponding low H_2 and D_2 concentrations measured online with the mass spectrometer.

2. Apparatus

Differences in vapor pressure[a] between H_2, HD and D_2 can be used to purify HD through a distillation process in a rectification column [7]. Successive liquefaction and vaporisation along the column will increase the concentration of the lightest element at the top of the column allowing, in a first step, the extraction of H_2 from HD and, in a second step, the extraction of pure HD from the remaining $HD - D_2$ mixture.

Our apparatus is made of a stainless steel column (measuring 30 cm in length and 5 cm in diameters, see Fig. 1) filled with a 20 stages Stedman packing. The cells are separated by a 1 cm thick epoxy ring. The Stedman cells are made of two semi-spherical stainless steel grid fixed on their side like a lens as it is shown on Fig. 1. The bottom of the column (the boiler) is a copper pot which contains as much as 10 moles of liquid HD and the top (the condenser) is connected to a cryogenerator cold head.

Equipment of the apparatus includes platinum and carbon temperature probes and two heaters fixed on both the condenser and the boiler. The two heaters can provide up to 6 Watts of input power. In addition, three capillaries are connected to the bottom, middle and top part of the column in order to extract gaz samples at different stages of the rectification process in the column.

Concentrations are measured by a commercial "MKS-Spectra Microvision Plus" quadrupole mass spectrometer having a limited mass range (only 1 to 6 mass unit). We have added an input gaz manifold designed to guarantee no mass segregation during gaz transport from the working pressure of the distillator (500 mbar) to the working pressure of the spectrometer

---

[a] At 18K, vapor pressure for H_2, HD and D_2 are respectively: 461.2 mbar, 235.2 mbar and 116.3 mbar, leading to a vapor pressure ratio of order 2 between HD and its impurities H_2 and D_2.

Figure 1. 3D view of the rectification column. From top to bottom: the condenser cryogenerator, the stainless steel column and the copper pot. We also see the three extraction capillaries. The column is filled with 20 Stedman cells as shown on the right.

(1.10^{-6} mbar). D_2 concentration can be measured down to 10^{-5} and H_2 down[b] to 2.10^{-4}.

A set of 10 empty tanks (22.8 liters each) are connected to the output of the distillator (from one of the three capillaries) via a mass flow controller in order to extrat the H_2 or HD rich samples. The extration flow ranges from 1 to 20 ml/min. Finally, an empty tank of 140 liters is connected to the column with a check valve (cracking pressure of 2.5 bar) for obvious security reason. The overall system is controlled by computer via a LabView interface.

3. Distillation procedure

We have distilled 6.2 moles of HD gaz containing 0.5% H_2 and 0.65% D_2 which is typical commercial "pure" HD. The control of the distillation is achieved by the two heaters providing the input power P_c (P_b) on the condenser (boiler). We control the reflux of liquid and vapor in the column by tuning P_b and the working temperature (or pressure) by controlling

---

[b]Limitation is due to dissociation of HD molecules leading to a significant background at mass 2 from D^+ ions.

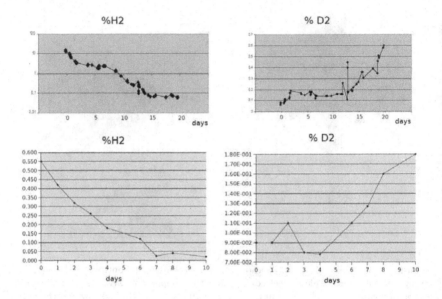

Figure 2. Evolution of H_2 and D_2 concentrations of extracted gaz samples. Top: for the first distillation with an initial gaz containing 0.5% H_2 and 0.65% D_2 (different extraction flow were tested ranging from 1 to 10 ml/min, days 2 to 6 correspond to 1 ml/min), note the logarithmic scale for the graph on the left. Bottom: for the second distillation of Sample II gaz (extraction flow equal to 1 ml/min).

P_c. We found that the best efficiency is given by the lowest boiling rate (P_b = 0W) and the lowest temperature. This can be understood by the fact that the equilibrium between liquid and vapor is reached very slowly in our distillator (time constant of the order of 6 hours) and that vapor pressure ratios increase when the temperature decreases (up to 2.76 at 15 K between H_2 and HD). Concentration of H_2 on top of the distillator was found to be 15.5% at 18 K. The Number of Theoritical Stages (NTS), as calculated with the Fenske relation, for a minimum reflux, is $NTS = 4.02$.

Condenser temperature was fixed to 18.25 K (291 mbars) with an H_2 rich sample on top of the distillator having the following concentrations: $[H_2] = 15.41\%$ and $[D_2] = 0.08\%$. This gaz was extracted to one of the output tanks with a flow maintained below 3 ml/min. H_2 and D_2 concentrations are monitored during the process (see Fig. 2). We ended up with three samples: 1.44 moles of H_2 rich gaz ($[H_2] = 2.46\%$, $[D_2] = 0.157\%$), 3.5 moles of H_2 poor gaz ($[H_2] = 0.08\%$, $[D_2] = 0.49\%$, Sample II) and 1.5 moles of D_2 rich gaz ($[H_2] = 0.06\%$, $[D_2] = 2.52\%$).

Sample II is then cryopumped back into the distillator for a second distillation. Concentrations of extracted gaz are shown on figure 2. We have three samples at the end of the distillation: the first one is suitable for BF polarization (BF Sample): 855 moles with $[H_2] = 0.26\%$ and $[D_2] = 0.09\%$, the second has been specifically produced for dynamic polarization (DNP sample): 350 mmoles with $[H_2] < 0.02\%$ reaching the detection limit of our spectrometer, and $[D_2] = 0.17\%$ and the rest of the gaz is $[H_2] = 0.12\%$ and $[D_2] = 0.5\%$. It is important to note that this distillation has been carried out with dynamic polarization in mind. The extraction strategy between the first and second distillation can significantly change the concentrations and the residual quantity of final samples. $[D_2]$ of the order of 0.08% (with $[H_2]$ still below 0.08%) would have been possible if Sample II were not chosen in the first place with such a low H_2 concentration. One should then perform different distillations for BF polarization (very low $[D_2]$) and proton or deuteron DNP (very low $[H_2]$ and $[D_2]$).

4. Relaxation time measurements

The DNP sample was then kept 1 week at 14 K before being transported to the Physics Department of the Rhur-Universität Bochum for a measurement of T_1^H. During the journey the gaz was at room temperature and at a pressure of 334 mbars for 10 hours.

At a temperature of 1.5 K, a field of 2 T and under radiation (β-source of 3.7 GBq [5]), T_1^H was measured to be 800 seconds. After 2.5 days of irradiation, T_1^H was still above 650 seconds. It is the first time that such a long relaxation time has been observed just after condensation of the HD gaining at least three orders of magnitude compared to previously available samples which contained typically 0.2% of H_2 and D_2.

References

1. S. Bouchigny et al., Nucl. Instr. and Meth. A 544 (2005) 417.
2. A. Honig, Phys. Rev. Lett. **19**, 1009 (1967).
3. T. Kageya, "Performances of frozen-spin polarized HD targets for Nucleon spin experiments", *This Conference*.
4. J.C. Solem, Nucl. Instr. and Meth. 117 (1974) 477.
5. E. Radtke et al., Nucl. Instr. and Meth. A 526 (2004) 168.
6. W. Meyer, "SPIN2004 Summary & Progress in DNP Targets", *This Conference*.
7. M. Rigney, et al. "Conversion, Measurement and Distillation of Hydrogen Isotopes: H_2, HD and D_2", *8th Int. Worshop on Polarized Target Materials and Techniques*, TRIUMPH, Vancouver, Canada, May 1996.

PERFORMANCE OF FROZEN-SPIN POLARIZED HD TARGETS FOR NUCLEON SPIN EXPERIMENTS *

T. KAGEYA[1,10], K. ARDASHEV[8], C. BADE[1,5], M. BLECHER[10],
A. CARACAPPA[1], A. D'ANGELO[7], A. D'ANGELO[7], R. DI SALVO[7],
A. FANTINI[7], C. GIBSON[1,8], H. GLÜCKLER[2], K. HICKS[5],
S. HOBLIT[1], A. HONIG[6], M. KHANDAKAR[4], S. KIZIGUL[5],
O. KISTNER[1], S. KUCUKER[9], A. LEHMANN[8], F. LINCOLN[1],
R. LINDGREN[9], M. LOWRY[1], M. LUCAS[5], J. MAHON[5],
L. MICELI[1], D. MORICCIANI[7], B. NORUM[9], M. PAP[2],
B. PREEDOM[8], A.M. SANDORFI[1], C. SCHAERF[7], H. SEYFARTH[2],
H. STRÖHER[2], C. THORN[1], K. WANG[9], X. WEI[1], C. WHISNANT[3]

*(1) Brookhaven National Lab., (2) Forschungszentrum Jülich GmbH,
(3) James Madison Univ., (4) Norfolk State Univ., (5) Ohio Univ.,
(6) Syracuse Univ., (7) Univ. di Roma, (8) Univ. of South Calolina,
(9) Univ. of Virginia, (10) Virginia Polytechnic Institute and State Univ.
E-mail: kageya@bnl.gov*

Frozen-spin HD targets have been used to measure beam-target double-polarization observables in pion photo-production at the LEGS (Laser Electron Gamma Source) facility. The HD targets are pure; the only unpolarizable nucleons associated with the target cell are sampled in empty-cell measurements and subtracted in the conventional way, resulting in exceptionally clean spectra. HD gas is isotopically purified by distillation and doped with a small amount of ortho-H_2 and para-D_2. It is then frozen into a mesh of pure aluminum wires which conduct away conversion heat from the decay of the H_2 and D_2 impurities. By holding the targets at low temperature and high field (12 mK and 15 Tesla) for about three months, most of the L=1 ortho-H_2 (para-D_2) converts to the magnetically inert L=0 para-H_2 (ortho-D_2) and results in a frozen-spin state for HD. Equilibrium polarizations for protons (H) and deuterons (D) of about 60 % and 15 %, respectively, have been obtained. H spins have been successfully transferred to D spins by saturating the H-D forbidden transition, which resulted in D polarizations of 31 %. Relaxation times of about 1 year for both H and D were measured during recent photo-production experiments.

---

*Supported by a research grant from the u.s. department of energy under contract no. de-ac02-98ch10886, the u.s. national science foundation and the istituto nazionale di ficisa nucleare

1. Frozen-spin HD

An H_2 molecule which consists of two identical protons has two spin states, with spins either antiparallel (para) or parallel (ortho). Ortho-H_2, which accounts for 3/4 of H_2 at room temperature, is readily polarized at high magnetic field and low temperature while the para-H_2 is not. At a low temperature, the ortho-H_2 decays to the para-H_2 state with a half life of about six days.

Due to a spin-spin coupling between an H in H_2 and an H in HD, a polarized ortho-H_2 can transfer its polarization to HD. A small concentration of polarized ortho-H_2 (on order of 10^{-4}) is used to polarize HD and in three months, most of the ortho-H_2 decays to the para state. Since there are no phonons to couple an S-wave HD to its crystal lattice, thus once polarized, HD has an extremely long relaxation time. D in HD can be polarized in a similar way, although to a lesser degree due to its smaller magnetic moment.

Target preparation begins by condensing and solidifying HD gas in a ^{4}He Production Dewar (PD). Thermal Equilibrium (TE) calibrations are carried out at 2 K in this dewar when relaxation times are still relatively short. The HD targets are then transferred to a Dilution Refrigerator (DR), polarized at low temperature and high field ($\approx$ 12 mK and 15 Tesla) and simply held in this state for about 3 months (aging) to reach to a frozen spin-state.

In November 2005, after three months of aging, a target was transferred back to the PD and polarizations were measured to be 60 % and 15 % for H and D, respectively. For target polarization measurements, a cross coil NMR with a phase sensitive detection has been used to detect interference signals between two coils by sweeping the magnetic fields [1].

The target was transfered to an In-Beam-Cryostat (IBC) cooled with a horizontal dilution refrigerator to a temperature of 250 mK with a holding field of 0.9 Tesla (maximum field is 1 Tesla). Transfer losses are larger for D and reduced the polarizations to 57% adn 8% for H and D, respectively. Polarizations were monitored frequently with NMR in the IBC every for both H and D. Relaxation times as long as two hundred days were observed for both D and H during photo-production experiments.

Pion production data were taken for spin orientations of both D and H parallel to the holding field for the first half of the run period; and H spins were flipped antiparallel to the field for the second half of the runs. This allowed spin-differences from reactions on H and D to be separated. H spins were flipped from + 57 % to about - 27 % with an allowed fast

RF transition by scanning the magnetic field at fixed frequency. (The inefficiency of the transition was due to an instability of the magnet power supply for IBC holding field.) After the run, the target was transfered to the PD and polarizations were remeasured, allowing an accurate interpolation of in-beam polarizations during the run. The average polarizations and relaxation times during the run are summarized in Table 1.

Table 1. Summary of average polarizations and relaxation times during recent runs at LEGS.

Period	Duration	P_D	P_H	$T_1(D)$	$T_1(H)$
Fall 2004	17 days	+ 8 %	+57 %, -27%	270 days	560 days
Spring 2005	32 days	+ 31 %	+30 %, -7%	350 days	220 days

In April 2006, another target was transfered to the IBC after about three months of aging. A saturated forbidden transition from H to D was performed in the IBC, increasing the D polarization to 31 %. The in-beam relaxiation times for D and H were measured as 350 days for D and 220 days for H. A summary of the average polarizations and relaxation times during this run is included in Table 1.

2. Preliminary spin asymmetries

Experiments were carried out using longitudinally polarized HD targets, circularly and linearly polarized photon beams and a nearly four π detector array for charged and neutral particles. The highly polarized photon beams were obtained by a laser-induced backward Compton scattering from electrons at the National Syncrotron Light Source located at Brookhaven National Laboratory.

The Kel-F target cell walls and the aluminum cooling wires are the only sources of unpolarizable nucleons and these are sampled in conventional empty-target runs. Figure 1 shows typical missing-energy distributions of the $\pi^0 n$ final state near the peak of the P_{33} delta resonance taken during the Spring 2005 run. The flux-weighted contributions from the target cell (about 20 %) are shown in the upper panels. The net yields from HD, obtained by subtracting this component, are shown in the lower panels. Flux-normalized yields collected with gamma-ray and deuteron spins parallel (antiparallel) are shown in the left (right) panels and show a pronounced asymmetry.

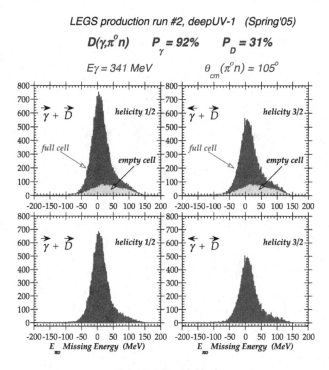

Figure 1. π^0n missing energy distributions for $E\gamma = 341$ MeV, $\Theta = 105$ degrees, $P_\gamma = 92$ % and $P_D = 31$ %. Flux-normalized full- and empty-cell contributions are shown in the upper panels and their difference in the lower panels. Yields with γ and D spins parallel (antiparallel) are plotted on the left (right). The equivalent helicity of the γ+n system is indicated.

References

1. A. Caracappa and C. Thorn, *Proceedings of 15th International Spin Symposium, Upton, New York.* 867 (2002).

Optical Pumping Method

A NEW ^{3}He POLARIZATION FOR FUNDAMENTAL NEUTRON PHYSICS[1]

Y. MASUDA, T. INO, S.C. JEONG, S. MUTO, Y. WATANABE
High Energy Accelerator Research Organization, 1-1 Oho, Tsukuba, Ibaraki 305-0801 Japan

V.R. SKOY[†]
Frank Laboratory of Neutron Physics, Joint Institute for Nuclear Research, 141980 Dubna, Moscow Region, Russia

A high ^{3}He nuclear polarization was obtained for a high-pressure ^{3}He gas in a sapphire cell. The phase-shift difference between ordinary and extraordinary rays is well controlled upon photon transmission through a birefringent window of sapphire in order to keep high circular polarization. The ^{3}He polarization was applied to a pulsed neutron Ramsey resonance. A Ramsey resonance was obtained as a function of the neutron time of flight.

1. Introduction

Currently, intensive work on the polarization of ^{3}He gas for a neutron spin polarizer is being carried out for neutron experiments in various fields. For example, space-time symmetry violation tests on neutron β decay, neutron transmissions through nuclear targets, and neutron-proton capture γ-ray. In these experiments, high ^{3}He polarization at high pressure and low background are very desirable. We have made a birefringent cell of sapphire for use with spin-exchange optical pumping. The phase-shift difference between ordinary and extraordinary rays is well controlled upon photon transmission through a birefringent window of sapphire so that high circular polarization is obtained in the cell. During the optical pumping, rubidium atoms, which are contained in a cell together with a ^{3}He gas, are polarized upon circularly polarized photon absorptions at a D1 resonance. The rubidium polarization may have a local distribution in the cell. The nuclear spin is then polarized up to an average rubidium polarization in the cell via hyperfine interactions upon atomic collisions, if there is no polarization relaxation [1]. The highest nuclear polarization will be obtained from a homogeneous distribution of 100%

---

[*] This work is supported by a research fund of Institute of Particle and Nuclear Studies, KEK.
[†] Work supported by grant of KEK foreign researcher.

rubidium polarization, where 100% photon polarization is required. Incident photon polarization, however, may decrease upon transmission through a window of the cell. This loss in the polarization recovers during propagation through polarized rubidium atoms by the spin-dependent absorption at the D1 resonance, but low rubidium polarization is produced near the window. In a birefringent cell of sapphire, the photon polarization can be controlled upon the transmission.

Sapphire has another advantage in the optical pumping. It is impervious to hot alkali metal vapor. This chemical property makes possible to use other alkali metal atoms, for example, potassium or sodium atoms at high temperature, which can speed up the spin exchange optical pumping process [2, 3].

For application to symmetry violation tests, sapphire has an advantage of very small neutron cross section. Neutrons with anti-parallel spins to the ^{3}He spins are preferentially absorbed because of a n-^{3}He bound state resonance, while neutrons with parallel spins are not absorbed. Therefore, neutrons are polarized upon transmission without neutron scattering, which is a dominant background source at the experiments. The neutron scattering cross section of ^{3}He is negligibly small. In addition, the sapphire cell can sustain a high pressure ^{3}He gas with flat windows, and then realizes high neutron polarization with a homogenous polarization distribution. This property is preferable for precision neutron experiments.

We have obtained a high ^{3}He polarization at a high pressure in the sapphire cell [4]. We are developing a new Ramsey resonance for the spin manipulation of pulsed neutrons by means of the ^{3}He polarization. A preliminary result was obtained.

2. Birefringent cell

We have made a sapphire cell, which is made from a cylinder of 30 mm inner diam, 46.5 mm length and 3 mm thickness, and two circular disks by diffusion bonding. The c axes of the three parts were aligned. A ^{3}He gas inlet port was attached to the cylinder. The inlet port comprises a sapphire pipe, several glass pipes, and a Pyrex pipe, which were connected by welding.

The photon polarization after transmission through the birefringent window of the ^{3}He cell is represented in terms of the phase-shift difference (θ) between ordinary and extraordinary rays. If the crystal c axis is in the plane of the window, the phase-shift difference is $\theta = 2\pi(n_o-n_e)l/\lambda$. Here, (n_o-n_e) is the difference between ordinary and extraordinary refractive indices, l is the window thickness and λ is the photon wavelength. At the rubidium D1 resonance, $\lambda = 795$ nm and $(n_o-n_e) = 0.00793$ for sapphire. Photon polarization P_{pho} after transmission through the window is estimated in terms of an initial photon state and a final photon state. If the initial photon state is completely polarized in a right-handed state, the photon polarization is represented as $P_{pho} = \cos^2(\theta/2) - \sin^2(\theta/2)$. The thickness of the window was $l = 3.014 \pm 0.001$ mm. The flatness of the window was $\lambda/4 \sim \lambda/10$. The incident photon polarization was 95%. Therefore, the phase shift difference upon transmission through the window is estimated to be $\theta = (23 \pm 4)°$ and then the photon polarization in the sapphire cell becomes $P_{pho} = 87\%$.

3. Optical pumping

The sapphire cell was placed in an oven to heat the rubidium, which was inserted into a 50 cm long and 12 cm diam solenoid. The magnetic field strength was 34 G with a homogeneity of 10^{-3} at the ^{3}He cell. We used a 95% linearly polarized laser beam of 11 W from a frequency narrowed laser diode array with a width of 0.25 nm for the spin-exchange optical pumping. The ^{3}He polarization (P_{He}) is estimated by a volume averaged rubidium polarization (P_{Rb}) in the cell, a rubidium-^{3}He spin exchange rate (Γ_{se}), and a ^{3}He polarization relaxation rate (Γ) as $P_{He} = P_{Rb}\, \Gamma_{se}/(\Gamma_{se} + \Gamma)$ [1]. The relaxation time of ^{3}He polarization ($1/\Gamma$) was measured at a room temperature by means of NMR to be $1/\Gamma = 24$ hours. The spin exchange rate depends on a rubidium atomic number density [5, 2], which depends on temperature [6]. The temperature of the cell was measured by a thermocouple, which was attached to the cylinder, to be 195°C. The temperature difference between the inside and outside of the cell is very small, because the thermal conductivity of sapphire is very high, which is 25 W/mK at 200 °C. If we use the rubidium temperature of 195°C and the Γ_{se} of Baranga et al. [2], $\Gamma_{se} = 1/5.0$ hours^{-1}, and then $\Gamma_{se}/(\Gamma_{se} + \Gamma) = 0.83$. Assuming the average rubidium polarization attains to the value of the circular polarization, the ^{3}He polarization is expected to be 72%.

4. ^{3}He polarization measurement

The ^{3}He polarization was measured by means of neutron transmission as a function of neutron energy. The ^{3}He cell together with the solenoid is placed in a pulsed neutron beam line at KEK. An incident-neutron monitor counter and a neutron beam collimator are placed upstream. Another neutron beam collimator and a neutron transmission counter are placed downstream. The neutron transmission was measured with the monitor and transmission counters as a function of the neutron time of flight, before switching on the laser power. The neutron transmission enhancement was measured after switching on the laser. The neutron cross section of polarized ^{3}He is represented as $\sigma_\pm = \sigma_0(1 \pm P_{He})$ for parallel and anti-parallel neutron spin states to the ^{3}He polarization [7]. σ_0 is the neutron cross section of unpolarized ^{3}He, which depends on the inverse of the neutron velocity $1/v$. If the incident neutron spin state is unpolarized, the neutron transmission of polarized ^{3}He nuclei (T) is represented as $T = A \exp(-\sigma_0 Nd) \cosh(P_{He}\sigma_0 Nd)$. Here, A is some constant, which is related to the incident neutron intensity and neutron attenuation by other materials. N is the ^{3}He nuclear number density and d is the thickness of the ^{3}He gas. The exponential attenuation factor, and thus the value of $\sigma_0 Nd$, is obtained from the neutron transmission for unpolarized ^{3}He (T_0). ^{3}He polarization is obtained from the enhancement of the neutron transmission (T/T_0), which is represented as $T/T_0 = \cosh(P_{He}\sigma_0 Nd)$. The neutron polarization after the transmission, which is represented as $P_n = \tanh(P_{He}\sigma_0 Nd)$, is obtained from the transmission enhancement as $P_n = \{1 - (T_0/T)^2\}^{1/2}$. The result of the ^{3}He polarization was 63 ±1 %. The ^{3}He pressure, which is related to N, was 3.1 atm. The neutron polarizations at cold and thermal neutron energies were very large, which were 97 and 90% with 15 and 22% transmission rates at 10 and 20 meV, respectively.

The present ^{3}He polarization obtained from the neutron transmission is lower than the estimated value. The rubidium polarization was not probably saturated by the frequency-narrowed laser beam of 11 W, because of a rather high spin destruction rate. The rubidium temperature was probably higher than expected because of laser heating [8].

5. Application to pulsed neutron Ramsey resonance

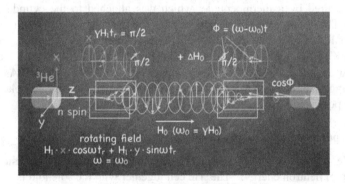

Figure 1. Pulsed neutron Ramsey resonance.

The principle of the pulsed neutron Ramsey resonance is shown in Fig. 1. The neutron spin after transmission through a polarized ^{3}He cell is rotated by $\pi/2$ in the fist radio frequency (RF) coils. During the rotation, the neutron spin is always vertical to the H_1 field which is a rotating field induced by the RF coils. After the first RF coils, neutron spin becomes vertical to the static field H_0. After Larmor precession in the H_0 field, the neutron spin enters the second RF coils. If the rotating field induced by the second coils is coherent with the first rotating field, the neutron spin continues the rotation about H_1 field, and then rotates by π in total after the second RF coils. The neutron spin direction is measured by means of the second polarized ^{3}He cell. If we modulate the phase of the second rotating field by ϕ, a $\cos\phi$ component is measured. An experimental result for the phase modulation as a function of the neutron time of flight is shown in Fig. 2. By using this method, we can control the neutron spin direction between the first and second RF coils. We can also measure the neutron velocity [9].

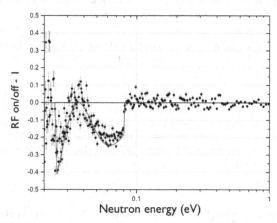

Figure 2. Ramsey resonance as a function of neutron time of flight. The phase of the second rotating field was modulated as a function of the neutron time of flight. The solid curve is a neutron transmission expected.

Acknowledgments

We would like to thank Prof. M. Kobayashi and Prof. F. Takasaki for their warm encouragement.

References

1. T. Chupp et al., *Phys. Rev.* **C36**, 2247 (1987).
2. A. Baranga et al., *Phys. Rev. Lett.* **80**, 2801 (1998).
3. E. Babcock et al., *Phys. Rev. Lett.* **91**, 123003 (2003).
4. Y. Masuda et al., *Appl. Phys. Lett.* **87**, 053506 (2005).
5. B. Larson et al., *Phys. Rev.* **A44**, 3108(1991).
6. T. Killian, *Phys. Rev.* **27**, 578 (1926).
7. M. Rose, *Phys. Rev.* **75**, 213 (1949).
8. D. Walter et al., *Phys. Rev. Lett.* **86**, 3264 (2001).
9. Y. Masuda et al., *Physica* **B356**, 182 (2005).

THE MIT LASER-DRIVEN TARGET OF NUCLEAR POLARIZED HYDROGEN GAS

B. CLASIE[1], C. CRAWFORD[1,*] D. DUTTA[2], H. GAO[1,2], J. SEELY[1]

W. XU[1,2,†]

[1] *Laboratory for Nuclear Science, Massachusetts Institute of Technology, Cambridge, MA 02139, USA*

[2] *Triangle Universities Nuclear Laboratory, Duke University, Durham, NC 27708, USA*

The laser-driven target at the Massachusetts Institute of Technology (MIT) produced nuclear polarized hydrogen gas in a configuration similar to that used in scattering experiments. The best result achieved was 50.5% polarization with 58.2% degree of dissociation of the sample beam exiting the storage cell at a hydrogen flow rate of 1.1×10^{18} atoms/s.

Laser-Driven Sources (LDS) produce nuclear polarized Hydrogen (H) or Deuterium (D) gas that can be fed into a storage cell. The combination of an LDS and a storage cell is called a Laser-Driven Target (LDT). Nuclear polarization of the target gas is achieved by spin-exchange collisions with optically pumped alkali atoms.

The valence electron of an alkali, in this case, potassium, can be polarized through optical pumping in a magnetic field of ~1 kG using circularly polarized laser light. The valence electron of potassium is excited from $4^2S_{1/2}$ to $4^2P_{1/2}$ followed by relaxation back to the $4^2S_{1/2}$ state. The helicity and wavelength of the laser are chosen to excite one of the two magnetic substates, resulting in depopulation of that substate and polarization of the potassium species.

Spin exchange collisions transfer the polarization from the potassium species to the H or D electron and the hyperfine interaction during H-H

---

*Present address: University of Tennessee, Knoxville, Tennessee

†Present address: Shanghai Institute of Applied Physics, Chinese Academy of Science, Shanghai, P.R. China

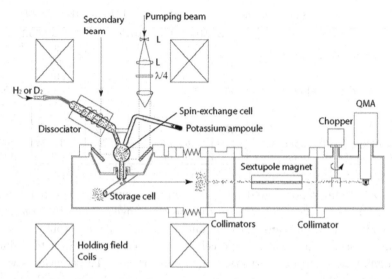

Figure 1. Laser driven target setup. Note that, for clarity, the polarimeter arm, storage cell, dissociator, and potassium ampoule are shown rotated by 90° from their actual positions.

or D-D collisions transfers the electron spin to the nucleus[1,2]. If there are many H-H or D-D collisions, the rate of transfer of spin to the nucleus equals the reverse rate, and the system is in Spin-Temperature Equilibrium (STE)[3]. LDSs are designed with the dwell time constant much greater than the STE time constant to guarantee H or D nuclear polarization. Moreover, STE has been verified in laser-driven sources and targets[4,5,6].

Production of nuclear polarized H or D atoms for use in internal targets is traditionally achieved using Atomic Beam Sources (ABS). The typical ABS flux, at present, is 6×10^{16} atoms/s. The typical flux from a Laser-Driven Source (LDS) is greater than 1×10^{18} atoms/s, compensating for the smaller polarization, and allowing for a potentially higher Figure-Of-Merit (FOM). Furthermore, LDSs offer a more compact design than ABSs.

Figure 1 is a schematic view of the MIT LDT. The best FOM from the target was recently reported in a Rapid Communication[7]. Polarized H is the first priority of this target and the source geometry was optimized for H. Molecular H was flowed into a dissociator at $1 - 2 \times 10^{18}$ atoms/s and the molecules were dissociated into atoms by a strong RF field. The atoms flowed through a capillary at the bottom of the dissociator and into the spin-exchange cell. Attached to the spin exchange cell by another capillary was a side arm where a potassium ampoule was heated to $200 - 250°C$

Table 1. FOM results from the HERMES ABS, IUCF LDT and the MIT LDT.

	HERMES (ABS)		IUCF (LDT)		MIT LDT Original	MIT LDT Large-1
Gas	H	D	H	D	H	H
Flow (F) (10^{16} atoms/s)	6.57	5.15	100	72	110	110
Thickness (t) (10^{13} atoms/cm^2)	11	[10.5]	50	50	150	150
f_α			~0.48	~0.48	0.56	0.58
P_e			~0.45	~0.45	0.37	0.50
$\langle p_z \rangle$	0.78	0.85	0.145	0.102	[0.175]	[0.247]
FOM F×$\langle p_z \rangle^2$ (10^{16} atoms/s)	4.0	3.8	2.1	0.75	3.4	6.7
FOM t×$\langle p_z \rangle^2$ (10^{13} atoms/cm^2)	6.7	7.6	1.1	0.52	4.6	9.2

and potassium vapor flowed into the spin-exchange cell. The potassium vapor and H gas were polarized in the spin-exchange cell and the polarized atoms flowed into a transport tube and a storage cell. The spin-exchange cell, transport tube and storage cell were coated with drifilm to reduce recombination and depolarization on these surfaces. These surfaces were heated to 200-250°C to prevent potassium from condensing on the walls, which would degrade the drifilm coating.

The storage (target) cell was 400 mm long and 12.5 mm in diameter. An electron polarimeter could be precisely aligned with metal bellows to sampling holes along the storage cell length. The gas entered at the top of the storage cell and exited from the sampling holes in the side of the cell or from the cell ends. The 90° angle between the entrance and sampling holes ensured that atoms underwent at least one wall collision in the storage cell before entering the polarimeter. The results from two spin-exchange cells, "Original" and "Large-1", are reported herein. The Original spin-exchange cell design was spherical with an inner diameter of 4.8 cm. Large-1 was a cylindrical cell optimized by a Monte Carlo simulation described below.

The best results achieved by the MIT LDT are shown in Table 1 together with results from the HERMES ABS[8] and the IUCF LDT[6]. For the MIT LDT the degree of dissociation of atoms exiting the storage cell sampling hole (f_α) and the electron polarization of atoms (P_e) was used to determine the density averaged nuclear polarization ($\langle p_z \rangle$), given by

$$\langle p_z \rangle = \frac{f_\alpha P_e}{f_\alpha + \sqrt{2}(1 - f_\alpha)}.$$

A Monte Carlo simulation of spin-exchange optical pumping was developed for this target. In the simulation, a hydrogen atom moved ballistically between wall collisions in the spin-exchange cell, transport tube and storage cell. The hydrogen atom could recombine at a wall collision with the probability proportional to the density of atomic hydrogen at the surface. Spin-exchange collisions were included by allowing the hydrogen atom to interact with the average potassium electron polarization and hydrogen electron and nuclear polarization, which were functions of position in the apparatus. Free parameters in the simulation were determined by fitting the Monte Carlo results to the experimental results from the Original configuration shown in Table 1.

The simulation was used to optimize the spin-exchange cell and transport tube geometry within the constraints of the existing vacuum chamber. The predicted results from the optimized design (called "Large-1") were $f_\alpha = 51\%$ and $P_e = 57\%$ at a flow rate of 1.1×10^{18} atoms/s. These results were in good agreement with the experimental results for Large-1 shown in Table 1. Furthermore, the simulation predicted an additional 20% could be achieved in the FOM with the design of a new vacuum chamber and the use of a cylindrical spin-exchange cell with a larger volume.

Polarized hydrogen gas has been produced at a flow rate of 1.1×10^{18} atoms/s, where the sample beam exiting the storage cell had 58.2% degree of dissociation and 50.5% polarization. This represents a significantly better FOM than the HERMES ABS and the IUCF LDT. A Monte Carlo simulation of spin-exchange optical pumping was developed and used to optimize the dimensions of the apparatus and the predicted results were in good agreement with the experimental results.

We thank Tom Wise and Willy Haeberli for the construction of the storage cells; Michael Grossman and George Sechen for their technical support; J. Stewart and P. Lenisa for the information on the HERMES ABS target. This work is supported in part by the U.S. Department of Energy under contract number DE-FC02-94ER40818. H.G. acknowledges the support of an Outstanding Junior Faculty Investigator Award from the DOE.

References

1. W. Happer, Rev. Mod. Phys. 44 (1972) 169.
2. J. Wilbert, Doct. Thesis, University of Erlangen (2002).
3. T. Walker and L.W. Anderson, Nucl. Instr. And Meth. A 334 (1993) 313.
4. J. Stenger, K. Rith, Nucl. Instr. And Meth. A 361 (1995) 60.
5. J. A. Fedchak et al., Nucl. Instr. And Meth. A 417 (1998) 182.
6. R. V. Cadman, Doct. Thesis, University of Illinois at Urbana-Champaign (2001).
7. B. Clasie et al., accepted by Phys. Rev. A. as Rapid Communication.
8. A. Airapetian et al., Nucl. Instrum. Methods A (2005) 540.

A PROPOSAL OF POLARIZED ^{3}HE^{++} ION SOURCE WITH PENNING IONIZER FOR JINR

N.N. AGAPOV, N.A. BAZHANOV, YU.N. FILATOV, V.V. FIMUSHKIN,
L.V. KUTUZOVA, V.A. MIKHAILOV, YU.A. PLIS, YU.V. PROKOFICHEV,
V.P. VADEEV
Joint Institute for Nuclear Research, Joliot-Curie 6, 141980 Dubna, Moscow region, Russia
E-mail: post@jinr.ru

A polarized ^{3}He^{++} beam can be accelerated in NUCLOTRON. This gives an opportunity to study the feasibility of ^{3}He^{++} production by ionization of polarized ^{3}He gas in the Penning ionizer using of an ion trap and pulse extraction. The ^{3}He gas can be polarized by the technique of Rb-^{3}He spin exchange optical pumping. The expected intensity of polarized (up to 50%) ^{3}He^{++} ions can be 5×10^{11} electronic charge/pulse.

1. Introduction

The goal is to make an intensive polarized ^{3}He^{++} beam in the JINR superconducting accelerator NUCLOTRON. The installation of a ^{3}He^{++} polarized ion source at the accelerator complex of JINR, Dubna, will allow to continue the spin physics experiments. The source will consist of an apparatus for polarizing of ^{3}He gas and ionizing of polarized ^{3}He atoms. There are two methods of the ^{3}He gas polarization, namely, by spin exchange with optically pumped Rb-vapour and by metastability-exchange optical pumping ^{3}He atoms in their 1s2s^{3}S$_1$ state. Our choice is the Rb-^{3}He method. We are going to construct our own equipment for ^{3}He polarizing (see Fig. 1). The Rb-^{3}He polarizer will include FAP-SYSTEM (COHERENT production) 25 W laser (a wavelength is 795 nm, line-width 2 nm), optical system, polarizing glass cell (^{3}He, Rb and N$_2$), cell baking and Rb distillation, refilling system (^{3}He-N$_2$ mixture), magnet coil system, ^{3}He system of cryosorption purification, a system of polarization measurements. Optical system involves polarizing cube, mirrors, lenses and λ/4-plates. It is supposed to use pyrex glass (Corning 7740) cells. The polarizing cell parameters: volume is 100 cm^3, ^{3}He pressure – 1.5-3 bar, cell temperature —

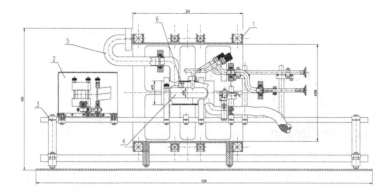

Figure 1. Rb-^{3}He polarizer. 1 — magnet system, 2 — optical system, 3 — support, 4 — polarizing glass cell containing ^{3}He, Rb and N_2, 5 — cell baking system, 6 — NMR-polarimeter coil.

180–200°C. The magnet system consisting of four coils will provide a 10 mT magnetic field with a degree of homogeneity 10^{-4}. The NMR-detection is the most important tool for monitoring the polarization process. This device will measure the signal of the forced precession of magnetization. The projected polarization of ^{3}He gas in the 0.1 l·bar cell is 50%. It is known the polarized gas can be moved from the polarizing field of Rb-^{3}He polarizer in a special container saving polarization. We would like to use this possibility having the special movable ^{3}He-cell and attach this one to ionizer. This is convenient under accelerator conditions. Then ^{3}He polarized beam is injected into an ionization volume restricted by Al-tubes drift system through a 0.1 mm capillary. The capillary gas flow is about 1 cm^3/hour or less.

2. Ionizer

The ionizer is based on the Electron Beam Ion Source (EBIS) in the reflex mode of operation processes and Penning ionizer [1]. Principle of operation is the ^{3}He^{++} ion accumulation in the ion trap with the following 10 μs pulse extraction [2] (see Fig. 2).

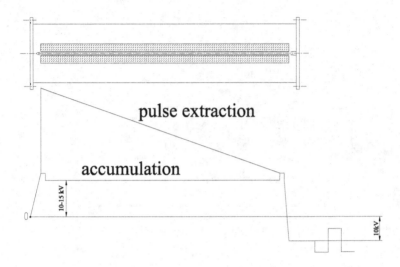

Figure 2. Diagram of the projected ionizer. The electrostatic voltage distribution on drift tubes at accumulation and extraction.

The ion trap is produced by space charge of oscillating electrons in a drift tubes region. In the same place the ionization by an electron impact is accomplished, also. In radial direction the electron cloud is confined by 1 T solenoid magnetic field. This field is needed to exclude ^{3}He depolarization on the walls of drift tubes. It is planned to use Nd-Fe-B permanent magnet for the ionizer.

Projected ionizer parameters:
energy of electrons – 10–20 keV,
effective current – 5–20 A,
length of ion trap – 1 m,
^{3}He^{++} ion beam intensity — 2×10^{11}–5×10^{11} electronic charge/pulse.
The advantages of this type of ionizer:
— high factor of ionization is not required, in other words, the ^{3}He holding time in the ion trap is short,
— ultrahigh vacuum is not required in comparison with EBIS ionizer.
It is scheduled to accumulate the volume charge of electrons in trap at 10–15 keV electron energy up to 5×10^{11}–10^{12} electronic charge and obtain ^{3}He^{++} polarized beam intensity up to 2×10^{11}–5×10^{11} electronic charge/pulse from the source.

The ionizer consists of a vacuum chamber, Nd-Fe-B permanent magnet system, an electron optical system (electron gun, electron optical structure and ion optics), an ionizator control system, 20 kV modulator, 10 µs system of a fast extraction and remote control system (Fig. 3). At the exit of the ionizer we have to install a spin-precessor for ^{3}He^{++} space spin orientation as the concording apparatus with direction of the accelerator magnetic field. The spin-precessor includes a spherical electrostatic mirror and solenoid.

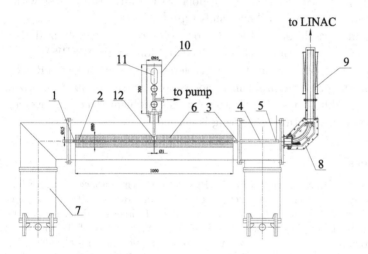

Figure 3. Schematic view of the ^{3}He^{++} polarized source. 1 — electron gun, 2 — drift tube sections, 3 — electron reflector, 4 — ion optical lens, 5 — movable Faraday cup, 6 — NdFeB permanent magnet, 7 — turbomolecular pump, 8 — spherical electrostatic mirror, 9 — spin precessor solenoid, 10 — flexible permanent magnet and magnetic shield, 11 — movable cell, 12 — capillary.

3. Depolarization effects

When an atom or ion has an unpaired electron, the depolarization can be induced by the hyperfine interactions between the electron and nucleus. However, if an external magnetic field B is much larger than a critical field B_c, proportional to the hyperfine interaction strength, the depolarization is reduced as a result of the decoupling of the nuclear spin from the electron spin. Under the external field B the primary nuclear polarization in the

^{3}He atom is reduced in the ^{3}He$^+$ ion [3] by the factor α:

$$\alpha = 1 - \frac{1}{2(1+x^2)}, \tag{1}$$

where $x = B/B_c$, with $B_c = -0.3096$ T for ^{3}He$^+$ ions. For $B = 1$ T $\alpha = 0.956$. But significant depolarization may be caused by the repetitive inelastic electron scatterings of ^{3}He$^+$ before ionization into ^{3}He^{++}. Also, de-ionization processes generated by collisions of produced ions with residual gases are dangerous [3]. The solution would be to increase the magnetic field. It is proposed to measure the polarization after the ionizer with a polarimeter, for example, a Lamb-shift type [4].

This ionizer could be used for the ionization of nuclear polarized ^{3}He$^+$ ions produced in the Spin Exchange Polarized Ion Source (SEPIS) [5].

As to the depolarization during acceleration in the NUCLOTRON [6], the estimates show this problem is not simple, there are 46 depolarizing resonances of different strength. To solve this problem some improvement of the magnetic structure of the ring is needed.

References

1. D. E. Donets, E. D. Donets et al., in: Proc. of Int. Symp. Space Charge Effects in Formation of Intense Low Energy Beams, Dubna, Russia, JINR, Feb. 15-17, 1999 (Eds: E.D. Donets, E.E. Donets and I.N. Meshkov) Dubna, 1999, p.99.
2. V. V. Fimushkin, Yu. K. Pilipenko, V. P. Vadeev, in: Proc. of the Int. Workshop Symmetries and Spin, Czech Republic, Prague, July 17-22, 2000 (Eds: M. Finger, O. V. Selyugin, M. Virius), *Czech. J. Phys.* **51** Suppl. A, A319 (2001).
3. M. Tanaka, N. Shimakura, Yu. A. Plis, *Nucl. Instr. & Meth.* **A 524**, 46 (2004).
4. Yu. A. Plis, in: Proc. of 13th Int. Symp. on High Energy Spin Physics, Protvino, Russia, Sept. 8-12, 1998 (Eds. N.E. Tyurin, V.L. Solovianov, S.M. Troshin and A.G. Ufimtsev) World Scientific, 1999, p.547.
5. M. Tanaka, Yu. A. Plis, E. D. Donets et al., *Nucl. Instr. & Meth.* **A 537**, 501 (2005).
6. I. B. Issinskii, A. D. Kovalenko, V. A. Mikhailov et al., in: Proc. VI Workshop on High Energy Spin Physics, Protvino, Russia, Sept. 18-23, 1995, Protvino, 1996, vol.2, p.207.

POLARIZED PROTON BEAMS IN RHIC

A.ZELENSKI, for the RHIC Spin Collaboration,
Brookhaven National Laboratory, Upton, NY.

Abstract. The polarized beam for RHIC is produced in the optically-pumped polarized H- ion source and then accelerated in linac to 200 MeV for strip-injection to Booster and further accelerated 24.3 GeV in AGS for injection in RHIC. A 50-55% polarization was routinely obtained at 100×100 GeV beam collisions in 2005 run. A maximum 65% polarization was measured in RHIC with bunch intensity reduced to $0.5 \cdot 10^{11}$ at injection energy. In the 2005 run the polarized beam was also accelerated to 205 Gev beam energy.

I. Introduction

Collisions of protons at energies $\sqrt{S} = 200$-500 GeV and transverse momentum $p_T > 10$ GeV /c are described as parton collisions (quarks, gluons), and for polarized proton beams these partons are polarized too. The analyzing powers for polarized parton scattering can be directly calculated in the frame of perturbative QCD. This provides a unique opportunity for the proton spin structure studies, fundamental tests of QCD predictions with possible extension to probe the physics beyond "Standard Model"/1,2 /. RHIC is the first collider where the "Siberian snake" technique was very successfully implemented to avoid the resonance depolarization during beam acceleration in AGS and RHIC / 3/. A luminosity of a $1.6 \cdot 10^{32}$ cm^{-2} sec^{-1} for polarized proton collisions in RHIC at up to $\sqrt{S} = 500$ GeV energy will be produced by colliding 120 bunches in each ring at $2 \cdot 10^{11}$ protons/bunch (see Fig.1).

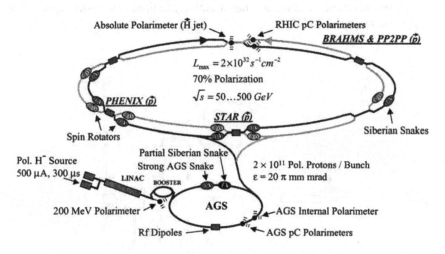

Fig.1.Accelerator –collider complex RHIC polarization hardware layout.

For the first time the intensity of the polarized beams produced in an optically pumped polarized H⁻ ion source (OPPIS) was sufficient to charge RHIC to the maximum intensity limited by the beam-beam interaction. Polarimetry is another essential component of the polarized collider facility. A complete set of polarimeters includes: a Lamb-shift polarimeter at the source energy, a 200 MeV polarimeter after the linac, and polarimeters in AGS and RHIC based on proton-Carbon scattering in Coulomb-Nuclear Interference region. A polarized hydrogen jet polarimeter was used for the absolute polarization measurements in RHIC/ 4/.

Longitudinally polarized beams for the STAR and PHENIX experiments are produced with spin rotators, which are tuned using "local polarimetrs" based on asymmetry in neutron production for pp collisions. The BRAHMS experiment uses transversely polarized beams. The STAR and PHENIX detectors provide complimentary coverage for the study of different polarization processes.

II. Polarized H⁻ ion source

The OPPIS technique is based on spin-transfer proton (or atomic hydrogen) collisions in an optically-pumped alkali metal cell. The modern technology involved – a superconducting solenoid, a 29.2 GHz microwave generator, and high power solid state tunable lasers – is essential to the OPPIS technique. In the BNL OPPIS, an ECR-type source produces a primary proton beam of a 2.0 – 3.0 keV energy, which is converted to electron-spin polarized H atoms by electron pick-up in an optically pumped Rb vapor cell (see Fig.2). A pulsed Cr:LISAF laser of a 1 kW peak at 500 us pulse duration is used for optical pumping. The nuclearly polarized H atoms are then negatively ionized in a Na-jet vapor cell to form nuclear polarized H⁻ ions.

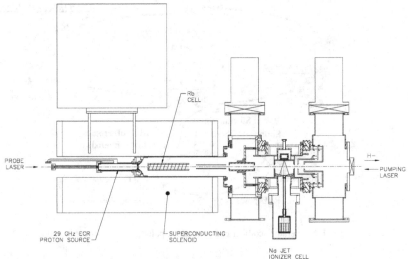

Fig.2. The RHIC OPPIS layout.

The RHIC OPPIS routinely produces 0.5-1.0 mA(maximum 1.6 mA) H⁻ ion current with 400 μs pulse duration and 80-82% polarization /5/. Polarized H⁻ ions are produced in the OPPIS at 35 keV beam energy. The beam is accelerated to 200 MeV with an RFQ and linac, for strip-injection to the Booster. About 60% of the OPPIS beam intensity can be accelerated to 200 MeV. The 400 μs H⁻ ion pulse is captured in a single Booster bunch (which contains about $4·10^{11}$ polarized protons). Single bunches are accelerated in the Booster to 2.5 GeV total energy and then transferred to the AGS, where they are accelerated to 24.3 GeV energy for injection to RHIC.

The OPPIS operates at 1 Hz repetition rate, and additional source pulses are directed to the 200 MeV p-Carbon polarimeter for continuous polarization monitoring by another pulsed bending magnet in the high-energy beam transport line. The p-carbon polarimeter was calibrated in comparison with the elastic p-deutron scattering, where the analyzing power was measured to better than 0.5% absolute accuracy at IUCF.

III. Polarized beam acceleration in AGS

Resonance beam depolarization occurs when the spin precession frequency is passing through specific resonance values (there are two types intrinsic and imperfection) during beam acceleration /5/. In the Booster, there are only two imperfection resonances, which can be corrected by harmonic coil correction. The AGS polarimeter at injection energy (Gg=4.5) is used for tuning of the harmonic coils. About 80 % beam polarization was measured at the AGS injection point in agreement with 200 MeV polarization measurements and depolarization in Booster is minimized to less than a few percent. There are many more imperfection and intrinsic resonances in the AGS. There is no space in the AGS structure for the full "Siberian snake" therefore a compromised solution was to use initially a partial "snake" which takes care of depolarization at imperfection resonances, and a pulsed dipole (AC-dipole) to reduce the depolarization at the four strongest intrinsic resonances. The AC-dipole introduce vertical beam modulations to enhance the betatron oscillation near the spin resonances to induce complete spin flips. This helps to cancel the main depolarization effects. The AC-dipole strength is not sufficient to flip the spin at the weak resonances and at one of the strong intrinsic resonance. Significant polarization losses were also introduced by the coupling resonances, which were introduced by the old solenoidal "partial snake". A new warm helical "partial snake", which was developed by M.Okamura (RIKEN) and installed for the 2004 polarized run eliminated these losses. In addition, after the OPPIS upgrade with a new superconducting solenoid, a higher current and a few percent higher polarization were obtained. The higher current allowed strong dynamical beam collimation in the Booster to reduce beam emittance in the AGS, which also helped to reduce resonant depolarization. As a result, a 65% beam polarization was recorded at the RHIC injection energy. This polarization was measured at the half of the regular beam intensity. The intensity dependence of depolarization at the intrinsic resonances was observed. It is likely caused by the space charge effect on the spread of the spin tune. At the regular beam intensity of a $1.0 \cdot 10^{11}$ p/bunch the polarization drops to 55-60%. About 15% residual depolarization was predicted from losses at the weak intrinsic resonances. To cancel all the resonance depolarization a new strong superconducting helical snake was developed and commissioned in 2005 run. Detailed calculations showed advantages in using a combination of both "warm" and superconducting snakes/6/. With the new snake the AC-dipoles are not required. This should eliminate the intensity limitation on polarization. The AGS goal for the 2006 run is to produce $1.5 \cdot 10^{11}$ bunch intensity at 60-65% polarization.

IV. Polarized beams in RHIC

The Relativistic Heavy Ion collider is the first high-energy machine where polarized proton accelearation was included in the primary design. There are two "Siberian snakes" in each ring to meet the conditions that the "snake" rotation is much larger than the total rotation from all other resonances up to highest 250 GeV beam energy. The RHIC "full Siberian snake", which rotates spin direction for 180° is a superconducting helical magnet system of about 10 m long. Two helical spin rotators in each ring produce the longitudinal polarization for experiments in STAR and PHENIX detectors.

Up to 120 beam bunches can be accelerated and stored in each ring. The polarization direction of every RHIC bunch is determined by the spin-flip control system in the polarized ion source. Every single source pulse is accelerated and becomes a RHIC bunch of the requested polarity. By loading selected patters of spin direction sequences in the rings (such as: +-+- in one ring and +--+ in another) the experiments have all possible spin directions combinations for colliding bunches which greatly enhance the systematic error control. Amazingly the spin direction of every bunch is kept under control during acceleration, storage time, and can be even measured in the CNI polarimeter (see Fig.3). There is also a plan to commission the AC-dipole "spin-flipper" to flip the spin direction of all bunches during the store for even better systematic error control. In the OPPIS the difference in polarization for opposite spin direction is less than 0.5% and spin –flip correlated beam current modulations are less than 0.01%. The OPPIS current pulses are obtained by chopping the dc current, therefore the current variations from pulse to pulse are very small. Due to large beam intensity losses during acceleration (typically only 10% of the primary source pulse current is stored in a RHIC bunch) there can be large variations of these losses which may cause significant bunch intensity variations (not correlated with the spin flip). These variations can be reduced to less than a few percents by the careful tuning in all accelerator and transfer lines.

A maximum polarized beam luminosity of a 10^{31} $cm^{-2} sec^{-1}$ (about $6 \cdot 10^{30}$ $cm^{-2} sec^{-1}$ averaged over 8 hrs store time) was obtained with 110 bunches of $0.9 \cdot 10^{11}$ bunch intensity in the 2005 run. The bunch intensity increase was limited by polarization losses in AGS and beam intensity losses in RHIC, caused by vacuum pressure rise and beam-beam interaction at three collision points. There was observed beam emttance degradation during the RHIC energy ramp which also reduced the luminosity. It is expected that the vacuum system upgrade and running just the STAR and PHENIX experiments will allow increased bunch intensity to $1.4 \cdot 10^{11}$ with higher polarization from AGS (due to two snake operation mode). A small beam emittance out of AGS was already demonstrated in the 2005 run, and with better control of beam injection to RHIC and energy ramp, the emittance degradation should be reduced. Therefore the optimistic projections for the 2006 run are about $2.5 \cdot 10^{31}$ cm^{-2} sec^{-1} for average store luminosity and 60% beam polarization.

V. Polarimetry

The proton-Carbon CNI polarimeters in AGS and RHIC are based on elastic proton scattering with low momentum transfer (Coulomb Nuclear Interference region) and measurement of asymmetry in recoil carbon nuclei production as described in detail elsewhere/7/. A very thin (30 nm thick 5 um wide) carbon strip in the high intensity circulating beam produces very high collision rates and a very efficient DAQ system acquires up to $5 \cdot 10^6$ carbon events /sec, which provides a unique bunch by bunch beam polarization measurements as shown in Fig. 3. Polarization measurement during the beam energy ramp was implemented in AGS and RHIC, to provide an understanding of polarization losses pattern. The carbon target width is much smaller than the beam size and polarization profile can be also measured.

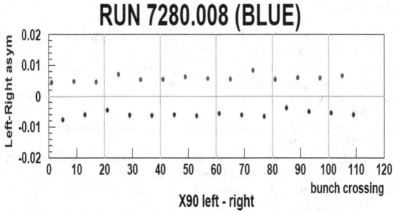

Fig.3. Polarization measurements in p-Carbon CNI polarimeter for a 28 bunch fill.

The absolute beam polarization at 100 GeV beam energy was measured with a polarized H-jet polarimeter which is also based on elastic proton-proton scattering in the CNI region. Due to particle identity, polarization of the accelerated proton beam can be directly expressed in terms of proton target polarization, which can be precisely measured by Breit-Rabi polarimeter. With the record polarized jet beam intensity of a $12.4 \cdot 10^{16}$ atoms/s obtained in this atomic beam source /4/, and with the expected beam intensity in Run-6 , a statistical error of about 8% can be obtained for polarization measurement in each RHIC store. The simultaneous measurements in p-Carbon and H-jet polarimeters provide the calibration for p-Carbon analyzing power. The fast pC polarimeter measures possible polarization degradation during the store duration. A precise measurement of pp analyzing power at 100 GeV was already extracted from the Run-4 data/8/. After analyzing power measurement at full 250 GeV RHIC beam energy it will be possible to use a higher thickness (of about 10^{15} atoms/cm^2) of unpolarized hydrogen jet target for precise absolute polarization measurements in RHIC.

VI. Conclusions

The RHIC spin program is a great beneficiary of the latest development in the polarized ion source and polarized internal target technology. For the first time the polarized proton beam intensity in the high-energy accelerator is not limited by the polarized source intensity. The great advances in polarized electron source development which were reported at this Workshop make feasible a future polarization facilities like eRHIC, a proposed polarized electron-proton collider with 10^{34} cm^{-2} sec^{-1} luminosity/9/.

The author would like to acknowledge a great team work of accelerator physicists in collaboration with RHIC experimental groups which made all above discussed progress possible.

References.

1. G.Bunce et al., "Prospects for Spin Physics in RHIC", Ann.Rev.Nucl.Part.Sci., **50**, p.525 , (2000).
2. N.Saito, "First resuts from RHIC Spin program ", Proc. 16th Int. Spin physics Symposium, p.58, World Scientific, (2005).
3. I.Alekseev et al., "Polarized proton collider at RHIC", NIM A499, p.392, (2003).
4. A.Zelenski et al.," Absolute polarized H-jet polarimeter for RHIC". NIM A536, p.248, (2005).
5. A.Zelenski et al., "Polarization optimization studies in the RHIC OPPIS", Proc. SPIN 2002, AIP Conf.Proc.675, p.881, (2003).
6. T.Roser et al., « Acceleration of polarized proton beam using multiple Siberian snakes ", ibid.2, p.687.
7. O.Jinnouchi et al., "RHIC pC CNI Polarimeter" Proc. SPIN 2002, AIP Conf.Proc.675, p.817, (2003). A.Bravar et al., "Spin dependence in elastic scattering in the CNI region", ibid.2,p.700.
8. H.Okada et al., Measurement of analyzing power in pp elastic scattering with the polarized H-jet target" to be published in Phys.Rev.Lett.
9. A.Deshpande, "Future colliders", Proc. 16th Int. Spin physics Symposium, p.192, World Scientific, (2005).

POLARIZED ³HE TARGETS AT MAMI-C*

JOCHEN KRIMMER

Institut für Physik, Johannes Gutenberg-Universität Mainz
Staudinger-Weg 7,
55128 Mainz, Germany
E-mail: krimmer@uni-mainz.de

Polarized ³He gas targets at MAMI-C offer new possibilities for double polarized experiments with real and virtual photons. For these experiments large amounts of highly polarized ³He are needed, which are provided by a MEOP polarizer with a polarization of more than 70% at a production rate of 2 bar·l/h.

1. Introduction and Motivation

The new acceleration stage of the Mainz microtron MAMI-C will come into operation in 2006. It will provide a continuous beam of unpolarized or polarized electrons with a maximum energy of 1.5 GeV. Polarized ³He serves as a substitute for a polarized neutron target and interesting new experiments can be performed in this energy range.

Double polarized photoabsorption data on the neutron from the GDH experiment at ELSA[1] show a resonant behaviour around 1 GeV which cannot be explained by existing multipole analyses[2]. It will be clarified with the Crystal Ball detector at MAMI, if this resonant behaviour comes from double pion partial reaction channels or if a modification of the helicity amplitudes is needed.

Quasi-elastic scattering of polarized electrons on polarized ³He can be used to extract the electric form factor of the neutron[3]. An approved proposal exists to measure G_{en} at $Q^2 = 1.5$ (GeV/c)^2, where only data from deuterium targets exist so far.

After this motivation the main part of this paper deals with the production of highly polarized ³He and the technical details of the target cells: Sec. 2 describes the principle and the performance of the ³He polarizer. In Sec. 3

---

*This work is supported by the Deutsche Forschungsgemeinschaft (SFB 443) the main relaxation processes are given. The design of target cells and results from material tests are presented in Sec. 4.

2. Polarization

The ^{3}He polarizer in Mainz operates according to the Metastability Exchange Optical Pumping (MEOP)[4,5] principle. The 2^3S_1 state is reached via a gas discharge at pressures of 0.8-1.0 mb and can then be optically pumped by circularly polarized laser light at 1083 nm[a]. Nuclear polarization of the 2^3S_1 state is transferred to unpolarized ground state atoms via collisions. The gas is compressed to the desired pressures of about 5 bar in an unmagnetic piston where less than 2% of the polarization is lost[6]. More than 70% target polarization can be obtained at a flux of 2 bar·l/h[7,8]. The T_1 relaxation of the target cells is measured by applying a static magnetic B_1 field additionally to the holding field B_0. After switching off the B_1 field the induced voltage in pickup coils is measured. This signal is proportional to the polarization in the cell. The flipping angle of less than 2° leads to a polarization loss of less than 0.02 % per measurement.

3. Relaxation

Several mechanisms reduce the polarization, where the total relaxation rate is the sum of the individual rates. Provided that the holding field is homogeneous enough[9] the relaxation time at higher pressures is limited by the dipole-dipole coupling of the ^{3}He atoms ($T_1^{dipole} = 817 \text{h}/p[\text{bar}]$[10]). In a charged particle beam the helium atoms can be ionized and ^{3}He$_2^+$ molecules are formed which destroy the polarization. This process can be suppressed by adding small amounts (10^{-3}) of N_2 as quenching gas. At an electron beam with 10 μA, T_1^{beam} is approximately 150-200 h[11].

A large contribution to the total relaxation rate comes from the interaction of the ^{3}He atoms with the container walls. It has been observed that the T_1 time of glass cells decreases drastically after an exposure to a large magnetic field (1.5 T) at a MRI tomograph. This effect is caused by ferromagnetic impurities, as the original T_1 time can be regained by a proper demagnetization[6]. Further wall interactions are adsorption and dissolution. The first can be diminshed by a coating with cesium, which reduces the sticking time due to its low adsorption energy. Dissolution can be minimized by using a dense glass, e.g. aluminosilicate or by a Cs-coating

---

[a]2*15 W fibre lasers, IGP Photonics: YLD-15-1083

which also closes the pores of a more porous glass, like quartz.
Theoretical interest in a quantitative understanding of the wall relaxation processes exists since long ago[12]. In the "historical" models paramagnetic impurities, e.g. iron, have been assumed to cause relaxation via dipole coupling. It has been observed, however, that glass containers made from a special iron free Supremax glass, and ordinary Supremax cells have similar relaxation rates. A new model, which takes into account Fermi-contact interaction with dangling bond type defects in the glass matrix, can describe the temperature dependence and the absolute values of relaxation rates with realistic parameters[13].

4. Target design

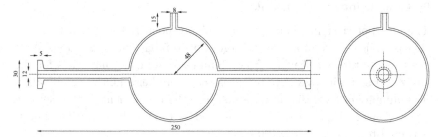

Figure 1. Sketch of a target cell used for electron scattering experiments. The numbers are given in mm.

A sketch of a target cell used for electron scattering experiments is given in Fig. 1. The electron beam runs in the horizontal direction. A prototype target cell has been manufactured from aluminosilicate glass (GE 180). Different materials of the entry and exit window for the electron beam, have been tested with this cell. The relaxation times that have been obtained from these measurements are given in Tab. 1. Only the last two lines show materials that can be used in a photon beam, here Mylar is preferred. From the materials that can be used in an electron beam titanium and diamond show the highest relaxation times, but both are not feasible for a real experiment. The diamond foils are not available in the quantities needed, and titanium produces too much background due to its large atomic number. The preferred solution is a beryllium foil which is covered by a thin aluminum foil. Owing to its low atomic number and its tensile strength beryllium is the ideal window material, but it is not ideal for keeping ^{3}He

Table 1. Relaxation times of different window materials.

material	T_1
blind flange	50 h
Copper (25 μm)	50 h
Diamond (30 μm)	70 h
Titanium (25 μm)	70 h
Beryllium (50 μm)	20 h
Beryllium (50 μm) + Al (10 μm)	50 h
Kapton (50 μm)	35 h
Mylar (50 μm)	45 h

polarized. A thin aluminum layer prevents the gas from direct contact with the beryllium.

5. Conclusions and outlook

The ^{3}He polarizer in Mainz provides highly polarized gas at a flux rate that is sufficient to fill a typical target cell to 5 bar within 2 hours. With a prototype target cell made from aluminosilicate glass, materials for the entry and exit windows have been tested, where beryllium in combination with aluminum gave the best results. For the experiment in the electron beam Cs-coated quartz cells will be used as their wall relaxation time is expected to be larger than 100 hours.

References

1. H. Dutz et. al., Phys. Rev. Lett. **94**, 162001 (2005).
2. D. Drechsel, S.S. Kamalov, and L. Tiator, Phys. Rev. **D63**, 114010 (2001), www.kph.uni-mainz.de/MAID.
3. J. Bermuth et. al., Phys. Lett. **B564**, 199 (1993).
4. F.D. Colegrove, L.D. Schearer, and G.K. Walters, Phys. Rev. **132**, 2561 (1963).
5. P.J. Nacher and M. Leduc J. Physique **46**, 2057 (1985).
6. J. Schmiedeskamp, Ph.D. thesis, Mainz (2005).
7. E. Otten, Europhysics News **35**, 16 (2004).
8. M. Wolf, Ph.D. thesis, Mainz (2004).
9. L.D. Schearer and G.K. Walters, Phys. Rev. **A139**, 1398 (1965).
10. N.R. Newbury, A.S. Barton, G.D. Cates, W. Happer, and H. Middleton, Phys. Rev. **A48**, 4411 (1993).
11. T.E. Chupp, R.A. Loveman, A.K. Thompson, A.M. Bernstein, and D.R. Tieger, Phys. Rev. **C45**, 915 (1992).
12. W. A. Fitzsimmons, L. L. Tankersley, and G. K. Walters Phys. Rev. **179**, 156 (1969).
13. J. Schmiedeskamp et. al., Eur. Phys. J. D, in press (2006).

DEVELOPMENT OF A POLARIZED ³HE TARGET AT RCNP

Y. SHIMIZU, K. HATANAKA, A.P. KOBUSHKIN[a], T. ADACHI[b],
K. FUJITA, K. ITOH[c], T. KAWABATA[d], T. KUDOH[e], H. MATSUBARA,
H. OHIRA[e], H. OKAMURA[f], K. SAGARA[e], Y. SAKEMI, Y. SASAMOTO[d],
Y. SHIMBARA, H.P. YOSHIDA[e], K. SUDA[d], Y. TAMESHIGE, A. TAMII,
M. TOMIYAMA[e], M. UCHIDA, T. UESAKA[d], T. WAKASA[e], AND
T. WAKUI[f]

Research Center for Nuclear Physics (RCNP) Osaka University, Ibaraki, Osaka 567-0047, Japan
[a]*N.N. Bogulyubov Institute for Theoretical Physics, 252143 Kiev, Ukraine*
[b]*Department of Physics, Osaka University, Toyonaka, Osaka 560-0043, Japan*
[c]*Department of Physics, Saitama University, Urawa, Saitama 338-8570, Japan*
[d]*Center for Nuclear Study (CNS), University of Tokyo, Wako, Saitama 351-0198, Japan*
[e]*Department of Physics, Kyushu University, Hakozaki, Fukuoka 812-8581, Japan*
[f]*Cyclotron and Radioisotope Center (CYRIC), Tohoku University, Sendai, Miyagi 980-8578, Japan*

A spin exchange type polarized ³He target was developed at RCNP. The ³He polarization was measured by the AFP-NMR method which provides a relative polarization value. In order to calibrate the absolute ³He polarization, we measured the spin correlation parameter C_{yy} of ³$\vec{\text{He}}(\vec{p}, \pi^+)$⁴He reaction. In the case $\frac{1}{2}^+ + \frac{1}{2}^+ \rightarrow 0^- + 0^+$, one can show from the parity conservation that the spin correlation parameter C_{yy} takes the constant value of +1. The present target had a maximum polarization of 0.19 and a thickness of 9.0×10^{19} cm^{-2}. With this target, we measured the differential cross section and the spin correlation parameter C_{yy} of $\vec{p}+$³$\vec{\text{He}}$ elastic backward scattering (EBS) at $E_p = 200$, 300, and 400 MeV. This process involves large momentum transfer and therefore a belief exists that EBS can provide an access to high momentum components of the wave function of the lightest nuclei.

1. Introduction

For several decades considerable efforts have been done to investigate structure of the lightest nuclei (d, ³He ⁴He) at short distances between constituent nucleons. Significant progress was achieved both in theory and experiment, first of all because high quality data on spin dependent observables were obtained with both hadronic and electromagnetic probes. Large part of these investigations consists of studies of elastic backward (in the center of mass system) proton-nucleus scattering (EBS). This process involves large momentum transfer and therefore a belief exists that EBS can provide an access to high momentum components of the wave function of the lightest nuclei. There are several cross section data of the $p + {}^3$He EBS, mainly at energies higher than 400 MeV. Analyzing powers were measured for the $p + {}^3$He elastic scattering at TRIUMF at 200 $\sim$ 500 MeV, but measurements were limited at relatively forward angles[1]. In order to measure the target-related spin observables, it was indispensable to develop a polarized target.

2. Polarized ^{3}He Target

A spin exchange type polarized ^{3}He target was developed at the Research Center for Nuclear Physics (RCNP), Osaka University. The schematic view of the major components of the target is shown in Fig 1. Two sets of coils with the axes orthogonal were used to change the direction of the polarization. One set of coils (main coil) generated a holding fields vertical to the direction of the beam, whereas the second set (RF coil) was used to adiabatically reverse the direction of the polarization transverse to the beam. In order to monitor the ^{3}He polarization, a small coil with the axis orthogonal to those of both the of the main and the RF coils was mounted on the target chamber (pick-up coil).

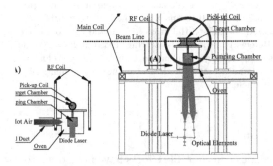

Figure 1. Schematic view of the polarized ^{3}He target system. The hatched part is the target chamber (upper), coupled to the pumping chamber (lower) by a thin drift tube.

The target cell has a double chamber structure consisting of two parts, *i.e.*, a target chamber and a pumping chamber, connected together by a thin transfer tube. By using two chambers we avoided the depolarizing effects of the proton beam on the Rb vapor. The target chamber had windows of 100 μm to reduce backgrounds from the scatterings on the windows. The target chamber was placed at the center of the main and the RF coils. The pumping chamber was placed in an oven. A hot air flow was used to heat the pumping chamber up to approximately 200 °C to get a sufficient Rb vapor density for optical pumping and spin exchange collisions. A high power diode laser and optical elements were introduced to polarize the Rb atoms in the pumping chamber. Polarized ^{3}He atoms were transferred to the target chamber by diffusion.

3. Measurement of the ^{3}He Polarization

The polarization of the target was measured by two independent methods. The first was a conventional NMR method coupled with the Adiabatic Fast Passage (AFP) method. The second method was the measurement of the spin correlation parameter C_{yy} of the $^3\vec{\text{He}}(\vec{p},\pi^+)^4$He reaction. The AFP-NMR was used to monitor the ^{3}He polarization regularly during the experiment and the second method was used to calibrate the NMR readings.

The second method is based on a priori known value of the spin correlation parameter C_{yy} of the $^3\vec{\text{He}}(\vec{p},\pi^+)^4$He reaction. For the special case of spins and parities of the initial and final states, $\frac{1}{2}^+ + \frac{1}{2}^+ \to 0^- + 0^+$, one can show from the parity conservation that the spin correlation parameter C_{yy} takes a constant value of unity[3].

The measurements were performed at RCNP. We used vertically polarized protons at incident energies of 300, and 400 MeV. The beam intensity was 10 to 40 nA, which was limited by counting rates. The proton polarization was about 70 %. The scattered π^+ particles were measured by the Grand Raiden spectrometer[2] to be set at 0°. In order to stop the beam and integrate the current a Faraday cup was installed inside of the first dipole magnet of the spectrometer. The targets labeled Cell#1 and Cell#2 were used for $E_p = 400$ and 300 MeV, respectively. The thicknesses of the Cell#1 and #2 were 5.3 and 6.1 mg/cm^2, respectively. The ^{3}He nuclear relaxation time of the Cell#1 and #2 were 20 and 15 hours, respectively.

The measured differential cross sections of $^3\vec{\text{He}}(\vec{p},\pi^+)^4$He reaction at 300 and 400 MeV is shown in the left panel of Fig. 2 together with existing data[4]. We see that our results are consistent with previous data. The ^{3}He

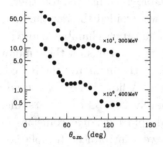 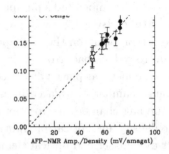

Figure 2. The differential cross section of the $^3\vec{\text{He}}(\vec{p},\pi^+)^4\text{He}$ reaction at $E_p = 300$ and 400 MeV (left) and the relation between the ^{3}He polarization and the AFP-NMR amplitude normalized to the value at amagat (right). The open and closed circles in the left panel show the present data and the previous data[4], respectively. The closed and open circles in the right panel show the results of Cell#1 and #2, respectively.

polarization obtained by the $^3\vec{\text{He}}(\vec{p},\pi^+)^4\text{He}$ reaction are shown in the right panel of Fig 2. The relation between the ^{3}He polarization P_{He} and the AFP-NMR amplitude normalized to the value at amagat [^{3}He] was determined as

$$P_{\text{He}} = (2.50 \pm 0.08) \times 10^{-3} \times V_{NMR}/[^3\text{He}]. \tag{1}$$

4. Summary

In order to perform the spin correlation measurements, we developed a spin exchange type polarized ^{3}He target. The absolute ^{3}He polarization was calibrated by the measurement of the spin correlation parameter C_{yy} of $^3\vec{\text{He}}(\vec{p},\pi^+)^4\text{He}$ reaction. The maximum and average values were about 19 % and 12 %, respectively. Because of inhomogeneity of the magnetic field at the experimental hall, the relaxation time decreased to 5 hours. In order to decrease the inhomogeneity of the magnetic field at the experimental hall, we will introduce the external Helmholtz coil.

References

1. D.K. Hasell et al., Phys. Rev. Lett. **74**, 502 (1986); E.J. Brash et al., Phys. Rev. C **52**, 807 (1995); R. Tacik et al., Phys. Rev. Lett. **63**, 1784 (1989);
2. M. Fujiwara et al., Nucl. Instrum. Meshodos Phys. Res. A **422**, 484 (1999).
3. G.G. Ohlsen, Rep. Prog. Phys. **35**, 760 (1972).
4. K.M. Furutani et al., Phys. Rev. C **50**, 1561 (1994).

DESIGN OF A POLARIZED ^{6}Li^{3+} ION SOURCE AND SIMULATIONS OF FEASIBILITY

A. TAMII, K. HATANAKA, H. OKAMURA, Y. SAKEMI, Y. SHIMIZU,
K. FUJITA, Y. TAMESHIGE AND H. MATSUBARA
Research Center for Nuclear Physics, Osaka Univ., Osaka 567-0047, Japan

T. UESAKA AND T. WAKUI
Center for Nuclear Study, Univ. of Tokyo, Tokyo 113-0033, Japan

T. WAKASA
Dep. of Physics, Kyushu Univ., Fukuoka 812-8581, Japan

T. NAKAGAWA
Institute of Physical and Chemical Research, Saitama 351-0198, Japan

Construction and feasibility test of a polarized ^{6}Li^{3+} ion source utilizing optical pumping and ECR ionizer is planned. Simulation of nuclear depolarization in the ECR ionizer is described.

1. Introduction

Nuclear spin-isospin excitations show rich and characteristic features in various nuclei. One of key points to study such excitations is to use probes which are selective for the reactions relevant to the physics of interest. In this paper we will report on design and simulation of a planned polarized ^{6}Li^{3+} ion source. Our primary interest is to study spin dipole resonances (SDRs) of various targets by measuring ($^6\vec{\text{Li}}$, ^{6}He) reaction at 100 MeV/A. For this purpose, production of ^{6}Li^{3+} ions and injection of them into the injector AVF cyclotron is essential.

Figure 1 shows concept of our polarized ^{6}Li^{3+} ion source (not to scale). A ^{6}Li atomic beam (~50 pµA) is produced by a lithium oven and collimators. ^{6}Li nuclei are polarized by the optical pumping method. We will use two diode lasers in Littrow design (Toptica Photonics DL100), each of which has a maximum power of 15 mW. The (F,M_F)=(3/2,3/2) state can be populated up to 95% by using the two lasers at 10 mW together with reflected light by a retro-reflector. The polarized atoms are injected into a superconducting ECR ionizer operated at 18 GHz after passing through an radio-frequency (RF) transition section and a polarimeter system. The atoms are ionized to 3+ by the ECR ionizer and are injected to the AVF cyclotron.

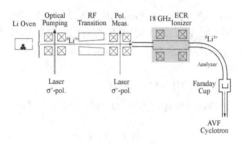

Figure 1. Schematic diagram showing the concept of a polarized ^{6}Li^{3+} ion source. The size does not scale.

2. Simulation of depolarization

A few depolarization processes in the ECR ionizer have been pointed out[?,?]: 1) depolarization in each elementary process in the plasma, e.g. ionization, charge-exchange (electron capture), and excitation of ions, 2) depolarization due to inhomogeneous magnetic field, 3) electron spin-flip due to electron spin resonance induced by RF power. In this section, we estimate the amount of depolarization in each process.

2.1. Depolarizaion in the ionization process

We assumed condition of ECR plasma as the one estimated from a basic study on the 14.5 GHz ECR ionizer at Univ. of Tsukuba (SHIVA)[2]: a uniform electron temperature of 580 eV with Boltzmann distribution, an ion temperature of 5 eV, a neutral gas density of 1.4×10^{10} cm^{-3}, and an electron density of 2.2×10^{11} cm^{-3}. Ionization rates, charge exchange rates, and atomic excitation rates were calculated from presently available parameters[3]. One of important numbers was confinement time (τ) of ions in the ECR plasma. We assumed τ=1 ms for ^{6}Li^{3+} based on the study of SHIVA and a proportional relation of τ to the ion charge. Amount of nuclear depolarization in each process was calculated from hyper-fine decoupling[4] in the averaged magnetic field and random spin orientation of captured and excited electrons.

After analytically solving a network calculation, we obtained a nuclear depolarization of 15% in the case of tensor polarization during the process of producing ^{6}Li^{3+} ions from ^{6}Li atoms. Thus 85% of the initial tensor polarization will remain. The result is plotted by the dashed line in Fig. 2 as a function of the assumed confinement time of ^{6}Li^{3+} ions.

2.2. Depolarizaion due to inhomogeneous magnetic field

When atoms or ions move in an inhomogeneous magnetic field, depolarization takes place. Only in the case of ^{6}Li$^{1+,3+}$ electron spin couples to singlet and nucleus feels the external magnetic field inhomogeneity and depolarizes.

This depolarization effect can be approximately calculated following Schearer et al.[5]. Assuming realistic magnetic field of an 18 GHz ECR and plasma conditions described above, we obtained nuclear spin relaxation time of 9 msec for ^{6}Li$^{1+,3+}$ ions. It corresponds to total depolarization of 10%. The result is plotted by the dotted line in Fig. 2.

2.3. Depolarizaion due to electron spin resonance

An RF field is applied to the ECR ion source for heating plasma. At the same place where the ECR occurs, electron SPIN resonance takes place and electron spin can be flipped. Since the resonance region is very thin, electron spin-flip is caused gradually and statistically. In the case of ^{6}Li^{2+} ions, this electron spin-flip can also cause nuclear spin-flip.

Statistical calculations of the electron spin-flip have been performed by assuming the same plasma conditions. The calculated average angle of the electron spin rotation in ^{6}Li^{2+} ions was only 3.5° during the confinement time of the ions. Thus the amount of electron spin-flip and nuclear depolarization is negligible (less than 0.2%).

2.4. Total depolarization and ionization efficiency

From the simulations we obtained that depolarizations caused by ionization processes and inhomogeneous magnetic field have non-negligible contributions. The calculated total depolarization for tensor polarization is plotted by the solid line in Fig. 2. In the present assumption of $\tau=1$ msec, the

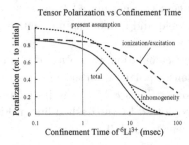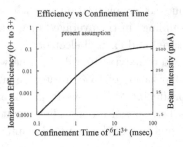

Figure 2. Result of simulation of depolarization effects for ionization/deionization/ excitation processes (dashed), inhomogeneous magnetic field (dotted) and total (solid). The right panel shows ionization efficiency of the ECR ionizer from ^{6}Li atoms to ^{6}Li^{3+} ions.

calculated depolarization is 25% (maintained polarization is 75%), which is not small but still acceptable. The depolarization can be decreased by somehow decreasing the confinement time.

3. Summary

We showed the concept of the planned ^{6}Li^{3+} ion source based on optical pumping and ECR ionizer. One of key points is amount of nuclear depolarization in the ECR ionizer. From simulations we obtained depolarization of 25% which is not small but still acceptable. Ionization efficiency of ^{6}Li atoms to ^{6}Li^{3+} ions in the 18 GHz ECR ion source estimated to be small. There may be compromise between the polarization and the beam intensity.

Acknowledgments

We are grateful to K.W. Kemper, E.G. Myers, B.G. Schmidt, and B. Roeder for valuable discussion.

References

1. T.B. Clegg et a., Nucl. Instrum. and Methods **A238**,195 (1985); M. Tanaka et a., Nucl. Instrum. and Methods **A524**, 46 (2004).
2. M. Imanaka, PhD thesis, Univ. of Tsukuba, unpublished.
3. G.S. Voronov, Atom. Data and Nucl. Data Tables **65**, 1 (1997); A. Muller and E. Saltzborn, Phys. Lett. A **62**, 391 (1977); D. Leep and A. Gallagher, Phys. Rev. A **10**, 1082 (1974); V.I. Fisher et al, Phys. Rev. A **55**, 329 (1997).
4. G.G. Ohlsen et al., Nucl. Instrum. Methods **73**, 45 (1969).
5. L.D. Schearer and G.K. Walters, Phys. Rev. **139**, A1398 (1965).

Nuclear Reaction Method

FOCAL PLANE POLARIMETER
FOR A TEST OF EPR PARADOX

K. YAKO, T. SAITO, H. SAKAI, H. KUBOKI, M. SASANO
Department of Physics, University of Tokyo, Bunkyo, Tokyo, 113-0033, Japan

T. KAWABATA, Y. MAEDA, K. SUDA, T. UESAKA
CNS, University of Tokyo, Bunkyo, Tokyo, 113-0033, Japan

T. IKEDA, K. ITOH
Department of Physics, Saitama University, Saitama, Saitama 338-8570, Japan

N. MATSUI, Y. SATOU
Department of Physics, Tokyo Institute of Technology,
Meguro, Tokyo 152-8551, Japan

K. SEKIGUCHI
Institute of Chemical Research (RIKEN), Wako, Saitama, 351-0198, Japan

H. MATSUBARA, A. TAMII
Research Center for Nuclear Physics, Osaka University,
Ibaraki, Osaka 567-0047, Japan

A proton polarimeter EPOL has been constructed at the focal plane of the spectrograph SMART at RIKEN. EPOL is designed for the test of EPR paradox in a two proton system by measuring the spin-correlation of the ^{1}H$(d, pp[^1S_0])n$ reaction at $E_d = 270$ MeV. EPOL consists of a spin-analyzer target (graphite slab), plastic scintillation counter hodoscopes and multi-wire drift chambers. EPOL is capable of identifying the trajectories of two protons before and after scattering in the analyzer target. EPOL has been calibrated at $E_p = 120$–160 MeV. The effective analyzing power is 0.17–0.23 and the figure of merit is 0.7–2.6 × 10^{-3}, both of which increase with incident energy.

In 1935, Einstein, Podolsky, and Rosen (EPR) claimed that quantum mechanics is incomplete in terms of local realism, which is a classical conception of nature[1].

This argument was developed by Bell quantitatively and he showed that any local hidden variable theory will result in an inequality, which contradicts quantum mechanical prediction[2]. Since the discovery of Bell's inequality, dozens of experiments have been performed by using entangled two-photon states to test the possibility of a local realistic theory, and most of them obtained results that the inequality is significantly violated. However, there has been only one experiment that tests Bell's inequality by means of a nuclear reaction[3]. Therefore we have started a series of experiments to test the inequality in a proton-proton system[4,5] and a proton-neutron system[6] by measuring the spin correlation. For this purpose, we have constructed a proton polarimeter (EPOL) at second focal plane (F2) of the spectrograph SMART[7].

The layout of SMART is shown in Fig. 1. The $^1S_0$ proton pairs are produced by the ^{1}H$(d,pp)n$ reaction at $E_d = 270$ MeV. The momentum acceptance of SMART at F2 is $\pm 2\%$ and the size of focal plane is 50 cm × 10 cm. Since the total energy of the proton pair is ~ 268 MeV, the positions of the two protons are symmetric with respect to the central ray in the dispersive plane. EPOL must have a two-proton tracking capability as well as a good figure of merit (FOM), defined as FOM = εA_C^2. Here ε and A_C are the double scattering efficiency and the effective analyzing power.

The polarimetry of EPOL is made by using the p+C inclusive scattering. Figure 2 shows a schematic view of EPOL. The proton pairs are detected by a multi wire drift chamber (MWDC1) and two layers of plastic scintillator arrays (HOD1) located upstream of the analyzer target. Each layer of HOD1 is horizontally segmented into eight scintillators for ease of event selection. The analyzer target is a graphite slab of 5 cm thick. The scattered protons pass through MWDC2 and MWDC3, and detected in two layers of plastic scintillator arrays (HOD2). 77% of the protons scattered in the analyzer target by angles less than 20° are in the acceptance of MWDC2, MWDC3, and HOD2.

MWDC1 has twelve layers of sensitive wire planes of a configuration of X-U-V-X'-U'-V'-X'-U'-V'-X-U-V. This redundancy of the sense planes enables us to reconstruct the two trajectories with high efficiency. The X', U', and V' planes are displaced by half of the cell size with respect to the X, U, and V planes, respectively, in order to solve the so-called left-right ambiguity. The configuration of MWDC3 is Y-Y'-X-X'-Y-Y'-X-X'. To retain the redundancy, MWDC2, which has two sensitive planes of U and V direction, has been installed. The half cell sizes are 10 mm for MWDC1, and 7-8 mm for MWDC2 and 3. The position resolution at each MWDC is 0.12–0.14 mm. The detection efficiency for each plane is more than 96–99%.

If two protons hit a cell or two adjacent cells in a MWDC plane, the position information is lost or becomes inaccurate and this often causes event loss or mistracking. Although x positions of the two protons are apart due to the kinematical condition, the chances of finding two protons in a y cell are quite high. To avoid mistracking, we put a software cut so that the y difference is larger than 20 mm in the analyzer target. Including the event loss of 30% by this cut, the efficiency of two-ray tracking is 0.55.

The effective analyzing powers of EPOL were studied at $E_p = 120$–160 MeV. Since polarized proton beams were unavailable at RIKEN, polarized protons were produced by induced polarization. The polarization of the protons produced at 19° by the $p + {}^{12}C$ elastic scattering at 160 MeV was expected to be 0.943 ± 0.013^8 due to time reversal invariance. Brass degraders were put in the downstream of the carbon target to obtain the protons with energies of 140 MeV, 130 MeV, and 120 MeV. The proton spin precessed around the vertical axis due to the magnetic field of SMART, whose total bending angle was 60°. The effective polarization along x-axis at EPOL was $P = 0.491$–0.522. Unpolarized proton beam was also directly transported to the EPOL and data were taken in order to eliminate the false asymmetry.

The up/down asymmetry is obtained by the sector method, where the neutrons with the scattering angles within the regions of $8.0° < \theta < 20.0°$ and $|\phi - \frac{\pi}{2}| < 66.8°$ ($|\phi - \frac{3\pi}{2}| < 66.8°$) are considered to be scattered in the upward (downward) direction (see Fig. 3). Figure 4 shows the result. The A_C value is 0.17–0.23 and increases with energy. It slightly depends on the position in F2, mainly due to the difference of the angular acceptance. The efficiency is $\varepsilon = 3$–5% and decreases with energy. The FOM value is 0.7–2.6×10^{-3}.

We note that the spin correlation measurement has been successfully carried out and that violation of Bell's inequality has been demonstrated by an accuracy of 2.9 standard deviations[5].

References

1. A. Einstein, B. Podolsky, and N. Rosen, *Phys. Rev.* **47**, 777 (1935).
2. J.S. Bell, *Physics* **1**, 195 (1964).
3. M. Lamehi-Rachti and W. Mittig, *Phys. Rev.* **D14**, 2543 (1976).
4. T. Saito et al., *Proc. the 19th European Conf. on Few-Body Problems in Physics, AIP Conf. Proc.* **768**, 62, (2005).
5. T. Saito, Doctral Dissertation (University of Tokyo, 2004).
6. S. Noji, K. Miki et al., in this Proceedings.
7. T. Ichihara et al., *Nucl. Phys.* **A569**, 287c (1994).
8. H.O. Meyer et al., *Phys. Rev.* **C27**, 459 (1983).

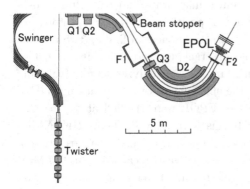

Figure 1. Layout of the spectrometer SMART.

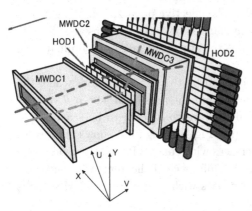

Figure 2. Schematic view of proton polarimeter EPOL.

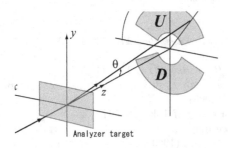

Figure 3. Sector definition. See text for details.

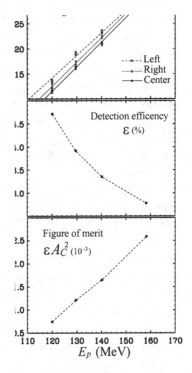

Figure 4. The effective analyzing powers, detection efficiencies ε, and figures of merit of EPOL.

DEUTERON BEAM POLARIMETRY AT NUCLOTRON*

V.P. LADYGIN, L.S. AZHGIREY, YU.V. GURCHIN, A.YU. ISUPOV,
M. JANEK, J.-T. KARACHUK, A.N. KHRENOV, A.S. KISELEV,
V.A. KIZKA, V.A. KRASNOV, A.N. LIVANOV, A.I. MALAKHOV,
V.F. PERESEDOV, YU.K. PILIPENKO, S.G. REZNIKOV, T.A. VASILIEV,
V.N. ZHMYROV, L.S. ZOLIN
Joint Institute for Nuclear Researches, Dubna, Russia

T. UESAKA, T. KAWABATA, Y. MAEDA, S. SAKAGUCHI, H. SAKAI,
Y. SASAMOTO, K. SUDA
Center for Nuclear Study, University of Tokyo, Tokyo, Japan

K. ITOH
Saitama University, Saitama, Japan

K. SEKIGUCHI
RIKEN (the Institute for Physical and Chemical Research), Saitama, Japan

I. TURZO
Institute of Physics Slovac Academy of Sciences, Bratislava, Slovakia

The current status of the deuteron beam polarimetry at Nuclotron is discussed. The preliminary results on the deuteron beam polarization measurements obtained with new tensor-vector polarimeter is reported.

1. Introduction

The base of the spin program at Nuclotron at the moment is the polarized ion source of atomic type POLARIS [1], which produces D^+ ions.

The polarization of the deuteron beam produced by POLARIS can be measured at the exit of LINAC by low energy polarimeters (LEP) [2], by

---

*This work is supported by the RFFR under grants No. 04-02-17107, 05-02-17743

several high energy polarimeters [3,4,5] after extraction of the beam from Nuclotron in the experimental hall and by new tensor-vector deuteron beam polarimeter [6] at Internal Target Station (ITS). The last one has been developed due to the installation plan of the atomic beam type polarized ion source CIPIOS to the LHE accelerator complex in order to establish a polarimetry suited for a vector-tensor mixed polarized beam.

2. Results on the deuteron beam vector polarization

The measurements of the extracted deuteron beam vector polarization has been performed in December 2002 at 5.0 and 3.5 GeV/c using polarimeter [4] based on the asymmetry measurements in quasi-elastic pp-scattering. The six-fold coincidences of counter signals from each pair of conjugated arms defined L or R scattering events.

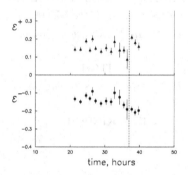

Figure 1. Left-right asymmetry of the extracted polarized deuteron beam at the Nuclotron at 5.0 and 3.5 GeV/c (last three points) versus time.

Polarimeter has been previously calibrated using Synchrophasotron polarized deuteron beam at 800 MeV/nucleon via $CH_2 - C$ subtraction using the world data on the analyzing power of free pp- scattering. The effective analyzing power A(CH_2) at forward proton scattering angle of 14° has been obtained at different energies and parametrized as a function of T_p. The systematic error in the measurement of the beam polarization is estimated as $\sim 5\%$.

Fig.1 shows the left-right asymmetries for "3-6" and "1-4" spin modes of POLARIS (ϵ^+ and ϵ^-, respectively) at the deuteron momenta 5.0 and 3.5 GeV/c versus the time of the measurements. The last 3 points in Fig.1 correspond to the measurements at 3.5 GeV/c. The averaged over spin modes polarization has been found to be 0.540 ± 0.019 and 0.606 ± 0.014 at 3.5 and 5.0 GeV/c, respectively [7]. The vector polarization of the beam obtained by LEP based on the d^4He- elastic scattering reaction at backward angles [2] was 0.59 ± 0.05. Good consistensy of the polarization values before and after acceleration and extraction demonstrate the absence of depolarization effects at Nuclotron.

3. Preliminary results with the tensorially polarized beam

The experiment on the calibration of new high energy polarimeter [6] based on dp- elastic scattering with tensorially polarized deuteron beam has been performed in June 05 using new ITS [8] at Nuclotron. The typical intensity of the beam in Nuclotron ring during the experiment was $2-3 \times 10^7$ deuterons per spill. The measurement of the beam polarization has been performed at 270 MeV, where well established data on analyzing powers exist [9].

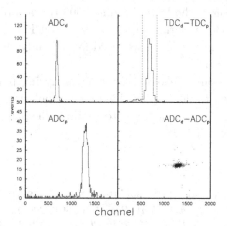

Figure 2. Typical ADC and TDC spectra for dp elasic events at 270 MeV.

48 scintillation counters based on Hamamatzu H7415 PMTs placed on the left, right, up and down were used at the same time. The detectors covered the angular range 60–130° in the center of mass. The typical size of the interacting beam was ±1 mm at 270 MeV.

The dp- elastic events at 270 MeV have been selected using the information on the energy losses in the plasic scintillators by protons and deuterons, their time-of-flight difference and interaction point position. The typical ADC and TDC spectra obtained at 270 MeV are shown in Fig.2. The contribution of background and accidential coincidences is negligibly small.

The polarization of the beam has been estimated from the part of the statistic at 270 MeV using the data on the analyzing powers from ref.[9]. The preliminary results on the tensor polarization of the beam are $0.623 \pm$

0.012 and −0.600 ± 0.009 for the "2-6" and "3-5" spin modes of POLARIS, respectively. The admixture of the vector polarization for these spin modes was found to be 0.207 ± 0.004 and 0.206 ± 0.003. The obtained values of the tensor polarization are in good agreement with the values obtained by LEP based on $^3He(d,p(0°))^4He$ reaction [2]: 0.69±0.13 and −0.67±0.16 for spin modes "2-6" and "3-5", respectively. The data analysis on the beam polarization values evaluation at 270 MeV is in progress.

4. Conclusions

Good consistensy of the vector polarization values of the deuteron beam before and after acceleration and extraction demonstrate the absence of depolarization effects at Nuclotron [7].

The polarization of the beam in June 2005 run has been measured by new tensor and vector polarimeter [6] at 270 MeV at ITS. This will allow to introduce the same polarization standard for 3 facility: RARF, Nuclotron and RIBF in the future.

New data have on dp elastic scattering analyzing powers up to 2000 MeV been obtained to develop the high energy polarimetry at Nuclotron and RIBF. Collaboration is planning to continue the data taking with new polarized ion source CIPIOS.

References

1. N.G. Anishchenko et al., *AIP Conf. Proc.* **95**, 445 (1983).
2. Yu.K. Pilipenko et al., *AIP Conf. Proc.* **570**, 801 (2001).
3. V.G. Ableev et al., *Nucl.Instr.Meth. in Phys.Res.* **A306**, 73 (1991).
4. L.S. Azhgirey et al., *Prib.Tech.Exp.***1**, 51 (1997); *Nucl.Instr.Meth. in Phys.Res.* **A497**, 340 (2003).
5. L.S. Zolin et al.,*JINR Rapid Comm.* **2[88]-98**, 27 (1998).
6. T. Uesaka et al., *JINR Preprint E1-2005-64*, Dubna (2005); to be published in Part.Nucl.Lett.; *CNS Annual Report 2004* **CNS-REP-66**, 81 (2005).
7. L.S. Azhgirey et al., *Part.Nucl.Lett.***2(125)** 91 (2005).
8. Yu.S. Anisimov et al., *Proc. of the 7-th Int. Workshop on Relativistic Nuclear Physics*, 25-30 August 2003, Stara Lesna, Slovak Republic, 117 (2004).
9. N. Sakamoto et al., *Phys.Lett.***B367**, 60 (1996); K. Suda et al., *AIP Conf. Proc.* **570**, 806 (2001); K. Sekiguchi et al., *Phys.Rev.***C65**, 034003 (2002).

DEUTERON POLARIMETER FOR ELECTRIC DIPOLE MOMENT SEARCH *

E. J. STEPHENSON[†]

Indiana University Cyclotron Facility
2401 Milo B. Sampson Lane
Bloomington, IN 47408, USA
E-mail: stephene@indiana.edu

We are developing a method to search for the electric dipole moment (EDM) of a charged particle using a storage ring. The test particles are injected into the ring as a polarized beam whose spin axis precesses (if there is an EDM) in the electric field that arises from $v \times B$ in the particle frame. We describe here the development of a continuously-operating polarimeter for a deuteron beam, which would potentially provide a sensitivity as low as 10^{-29} e·cm.

1. Introduction to the Storage Ring Method

An electric dipole moment (EDM) aligned along the spin axis of a particle violates parity (P) and time-reversal (T) symmetry. At present or proposed levels of experimental sensitivity, the Standard Model does not predict an EDM. But these levels encompass predictions by super-symmetric models needed to explain the matter–anti-matter asymmetry of the universe or to unify the fundamental forces. EDM searches on the deuteron using the methods described here are particularly sensitive to quark-based EDMs [1] and can reach statistical sensitivities of 10^{-29} e·cm.

Searches for an EDM involve the observation of changes to the particle precession rate when large external electric fields are applied. Searches using a beam in a storage ring offer the opportunity to expand the search to charged (and polarizable) particles and to use the much stronger electric field generated through $v \times B$ in the particle frame [2]. With a beam initially polarized along its momentum, an EDM would generate a vertical

---

*This work is partially supported by the US National Science Foundation under grant PHY-04-57219.
[†]for the Storage Ring EDM Collaboration, see http://www.bnl.gov/edm/

polarization component due to the precession caused by the **v×B** radial electric field. Any polarimeter for this search would need sensitivities as low as 10^{-6} and high efficiency.

The main obstacle for this method is the management of the much faster precession due to the particle's magnetic moment that takes place in the plane of the storage ring. Two methods have been studied.

The first is illustrated using the spin evolution equation:

$$\frac{d\vec{S}}{dt} = \frac{e}{m}\vec{S} \times \left[a\vec{B} + \left(\frac{1}{\gamma^2 - 1} - a \right) \vec{\beta} \times \frac{\vec{E}}{c} + \frac{\eta}{2} \left(\frac{\vec{E}}{c} + \vec{\beta} \times \vec{B} \right) \right] \quad (1)$$

where $a = (g-2)/2$ is the anomalous magnetic moment of the particle and η is the EDM. The precession created by $a\vec{B}$ may be cancelled by imposing a radial laboratory electric field so that the first two terms of Eq. (1) sum to zero [2]. This works well for the deuteron where a is small (-0.14) and leads to a design where a 3.5-MV electric field is used in a 13-m ring with a beam momentum of $p = 0.7$ MeV/c (or $T = 126$ MeV).

The second allows the precession caused by the magnetic moment to continue, but imposes a forced synchrotron oscillation at the same frequency as this precession. Then the EDM precession that accumulates for one part of this cycle is not completely cancelled by the other half, and the residual adds from cycle to cycle, eventually leading to a detectable signal [3].

2. Concept of the Polarimeter

The precession due to the EDM is allowed to accumulate as long as feasible, which is the time that the polarization can be maintained in this unstable equilibrium before it decoheres (about 100 s). The elimination of systematic errors demands that the polarization of the beam be monitored continuously during the beam store in order to demonstrate the time dependence expected of an EDM signal.

One way to obtain continuous measurements with high efficiency is to use an internal gas target to slow-extract the beam using Coulomb scattering. Particles removed from the beam continue downstream until they strike a thick target that acts as the polarization analyzer. The central opening in this annular target becomes the defining aperture for the ring. Scattering from the thick target to much larger angles carries the spin dependence needed for a good statistics measurement using a downstream array of detectors.

For both protons and deuterons, the most viable target material is carbon, since this material provides a large forward-angle peak in the analyzing power. The systematics for protons [4,5] and deuterons [6,7,8] have been investigated previously. Deuteron data between 70 and 175 MeV is lacking, so this energy region was investigated through new measurements.

3. Deuteron Elastic Scattering Data

Data were taken with the polarized deuteron beam at the KVI in Groningen, the Netherlands. For the first run at 76 and 113 MeV and at large scattering angles, the detectors were plastic scintillators for dE/dx and NaI assemblies for E. Detectors left and right were used to determine the asymmetry. The beam polarization was obtained from $d+p$ elastic scattering calibrated against Wit ała [9]. When it appeared that the most useful angles were smaller than those observed during the first run, additional measurements were made at 113 and 133 MeV using the BBS spectrometer.

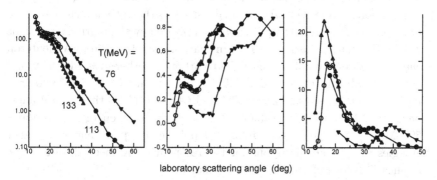

Figure 1. Measurements of the cross section, A_y, and figure of merit (different angle scale) for deuterons of 76, 113, and 133 MeV elastically scattering from carbon. The solid (open) points were measured with NaI (the BBS spectrometer).

These data are shown in Fig. 1 as angular distributions of cross section and the A_y analyzing power for deuteron elastic scattering. The relative performance of a polarimeter based on these measurements is given by the figure or merit, σA_y^2, which varies as $1/(\delta\epsilon)^2$ where $\delta\epsilon$ is the statistical error in the asymmetry that determines the EDM polarization.

Following an initial oscillation the cross section falls exponentially, indicating the dominance of scattering from the far side of the nucleus [10]. At the same angles, the analyzing power rises. These trends move to more forward angles as the energy rises.

4. Implications for an EDM Polarimeter

The studies with NaI detectors also measured inelastic deuterons and protons (mostly from breakup). Most of these reaction products have positive analyzing powers similar to those seen for elastic scattering, and so might prove useful if retained in the polarimeter acceptance. But the greatest impediment to good polarimeter performance is the large flux of breakup protons which have essentially no spin dependence. With expected detector rates of $10^6 - 10^7$/s, we anticipate that current mode detector readout might be required. So breakup protons must be eliminated ahead of the detector using range absorbers [7,8]. This leaves elastic scattering as the dominant contributor to the polarimeter detector rate.

The figure of merit shows peaks for foward angles ($\sim 17°$) at the higher energies and more backward angles ($\sim 40°$) at the lower energies that might be the basis for a polarimeter design. The advantage to the larger-angle choice is that the analyzing power would be larger, thus suppressing systematic errors in the polarization measurement. This energy range represents a transition from large angles to small angles as the optimum choice; above this energy the small-angle peak has been favored by all polarimeter designs [6,7,8]. Simulations using a 5.1 g/cm^2 carbon target at 126 MeV deuteron energy followed by a 1.3 g/cm^2 inert carbon absorber give efficiency and average analyzing power values of $\varepsilon = 0.18\%$ and $\langle A_y \rangle = 0.33$ for the backward-angle peak and $\varepsilon = 0.87\%$ and $\langle A_y \rangle = 0.11$ for the forward-angle peak. The forward-angle values continue to rise with energy up to about 500 MeV [6,7,8], leading to improved polarimeter performance. Higher energies may be advantageous for the resonance ring method [3] and the implications for polarimetry will be studied.

References

1. C.P. Liu and R.G.E. Timmermans, Phys. Rev. C **70**, 055501 (2004).
2. F.J.M. Farley *et al.*, Phys. Rev. Lett. **93**, 052001 (2004).
3. Y. Orlov *et al.*, in *6th Int. Conf. on Nucl. Phys. at Storage Rings*, Jülich-Bonn, Germany, 2005.
4. M.W. McNaughton *et al.*, Nucl. Instrum. Methods **A241**, 435 (1985).
5. S.M. Bowyer, Ph.D. thesis, Indiana Univ., 1994.
6. Y. Satou, DPOL polarimeter, private communication.
7. B. Bonin *et al.*, Nucl. Instrum. Methods **A288**, 389 (1990).
8. V.P. Ladygin *et al.*, Nucl. Instrum. Methods **A404**, 129 (1998).
9. H. Witała *et al.*, Few-Body Systems **15**, 67 (1993).
10. E.J. Stephenson *et al.*, Nucl. Phys. **A359**, 316 (1981).

Polarized Electron Production

127

OPERATION OF CEBAF PHOTOGUNS AT AVERAGE BEAM CURRENT > 1 mA[*]

MATT POELKER AND JOE GRAMES
Thomas Jefferson National Accelerator Facility, 12000 Jefferson Ave., Newport News, VA 23606, USA

> Photocathode operational lifetime of modern DC high voltage GaAs photoguns is limited primarily by ion backbombardment, the mechanism where residual gas at the cathode/anode gap is ionized by the extracted electron beam and back-accelerated toward the photocathode. Improving gun vacuum is an obvious way to prolong photocathode operating lifetime however this is not a trivial task. In this paper, the possibility of improving photocathode lifetime by merely increasing the laser beam diameter at the photocathode was explored. Lifetime measurements were made at beam currents from 1.5mA to 10mA using green laser light and a bulk GaAs photocathode inside a 100 kV DC high voltage load locked gun.

1. Introduction

Proposals for new electron/ion colliders, eRHIC [1] and ELIC [2], require exceptionally high average current polarized electron beam on the scale of tens of mAmperes, orders of magnitude above currents demonstrated at today's polarized electron accelerators worldwide. Proposals for new unpolarized electron accelerators such as the JLab 100kW free electron laser and the Cornell energy recovered linac [3] require even more current, on the order of ~ 100 mA, a significant extrapolation beyond the typical operating current of ~ 5mA demonstrated at today's JLab FEL. In addition, GaAs photoemission guns may find widespread use providing 10s to 100s of mAmperes for ion-beam cooling applications. All of these programs require better understanding of lifetime limitations of GaAs photoemission guns at high current. This submission describes the initial phase of a long-term program designed to study lifetime of GaAs photocathodes at beam current > 1mA.

Photocathode operational lifetime of modern DC high voltage GaAs photoguns is limited primarily by ion backbombardment, the mechanism where residual gas at the cathode/anode gap is ionized by the extracted electron beam and back-accelerated toward the photocathode. Ions with sufficient kinetic

---
[*] This work is supported by U.S. DOE under contract DE-AC05-84ER401050

energy damage the GaAs crystal structure or possibly sputter away the surface chemicals used to create the negative electron affinity condition. Improving gun vacuum is an obvious way to prolong photocathode operating lifetime because better vacuum means there will be fewer ions to damage the photocathode however, improving vacuum is not a trivial task. In this paper, the possibility of improving photocathode lifetime by merely increasing the laser beam diameter at the photocathode was explored. A very simple model suggests that for a large laser spot, total ion production at the cathode/anode gap remains the same, but ion damage is distributed over a larger area and therefore QE at specific photocathode locations degrades more slowly. Reality however is more complicated, in particular ion damage will not be uniformly distributed across the photocathode surface because ions created near the anode will be focused toward the electrostatic center of the photocathode and for gun configurations like those at CEBAF, one does not generally run beam from the electrostatic center of the wafer. Furthermore, the ionization cross section differs for each residual gas species and this cross section varies with electron beam energy, from 0V at the photocathode surface to 100kV at the anode. In addition, the stopping depth of each ion within the material varies as a function of gas species and energy. To date, the sensitivity of the photocathode QE to ions of different species and energies has not been measured.

2. Experiment

A 100 kV DC high voltage load-locked GaAs photogun was used to explore GaAs photocathode lifetime versus laser spot size (Fig.1). The 100 kV load loaded gun has been described at other workshops [4]. It consists of three vacuum chambers; load chamber, activation chamber and high voltage chamber with the best vacuum, 2×10^{-11} Torr, obtained in the high voltage chamber using a 120 L/s Perkin-Elmer sputter ion pump and three flange-mounted SAES GP500 NEG pumps. Beam leaves the gun through a large bore (2.5") NEG-coated beamtube toward a 15 degree bend magnet. The bend magnet allows illumination of the photocathode at normal incidence without using mirrors inside the vacuum chamber. There are four focusing solenoid magnets to manage the beam envelope and numerous steering magnets to keep beam centered in the beampipe. Beam is dumped at a tapered Faraday cup approximately 5 m from the photocathode. The Faraday cup was degassed at 450C for 24 hours prior to these measurements to limit outgassing during high current operation. In addition, two differential pump stations provide a factor of ~ 100 dump-vacuum isolation.

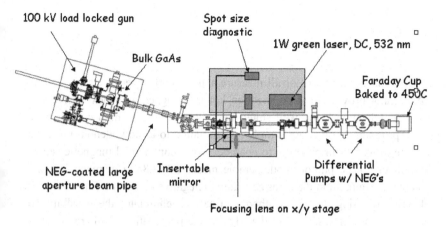

Figure 1. The 100 kV DC high voltage load-locked GaAs photogun and beamline.

Green light at 532 nm from a frequency-doubled Nd:YVO$_4$ laser (Coherent Verdi-10) was directed into the vacuum chamber through a vacuum window located near the bend magnet. The drive laser emits DC light, with maximum power set to ~ 1 W for these measurements. The amount of light at the photocathode could be varied with a computer-controlled attenuator that consisted of a fixed linear polarizer and rotatable halfwave plate. A 2 m focal length lens placed near the vacuum window of the beamline provided the tightest beam waist at the photocathode ~ 350 um diameter, FWHM. Larger laser spots were obtained by replacing the 2 m lens with a 1.5 m focal length lens positioned at locations on the laser table to provide the desired spot size at the photocathode. This simple manner of changing the laser spot size, where the laser beam passes through a beamwaist in front of the photocathode and expands to the desired spot size, produced clean Gaussian-shaped spots as determined using a commercial ccd camera and scanning razor-blade. The focusing lens was mounted to an x/y stepper motor stage to move the laser spot to different locations on the photocathode and to map photocathode QE before and after runs. The mirrors, waveplates, polarizers, etc. were all purchased with appropriate coating for green light.

Bulk GaAs was used for all measurements reported here. Of course bulk GaAs cannot provide high polarization but it is rugged and inexpensive with very high QE at 532 nm, well suited for this phase of the high current beam studies program. The GaAs sample was mounted to a molybdenum puck that could be moved from chamber to chamber using commercial magnetic sample manipulators. The photocathode sample measures 12.8mm diameter but only the center portion, 5mm dia., was activated to negative electron affinity using Cs and NF_3. Limiting the photocathode active area eliminates the possibility of inadvertent photoemission from the edge of the photocathode, where electrons travel extreme trajectories that do not terminate at the beam dump. Selective activation of the center of the photocathode was accomplished using a mask inside the activation vacuum chamber, a method similar to that employed at the Mainz Microtron [5].

The experiment consisted of numerous runs at different beam currents (1.5mA, 5mA, 7.5mA or 10mA) and with different laser spot sizes (342um, 842um or 1538um). Beam current, drive laser power and ion pump current at seven locations along the beamline were monitored for each run. The ion pump current provides a measure of vacuum along the beamline, and could be monitored to a very sensitive level using homebuilt ion pump power supplies with sub-nanoA current resolution [6]. Beam current was maintained constant throughout each run using a software feedback look that varied the amount of laser power delivered to the photocathode. The run might take minutes or hours, depending on photocathode lifetime for each configuration of beam current and laser spot size. Charge lifetime was defined as the amount of charge that could be extracted from the photocathode until the QE dropped to 1/e of the initial starting value. QE degradation across the surface of the photocathode could be monitored following each run by extracting ~ 1uA from the photocathode biased at -200V while scanning the laser across the photocathode using the x/y stepper motor stage. Figure 2 shows two QE scans: a) photocathode with relatively uniform high QE following activation and, b) after running ~ 100C of beam from one location. Note the appearance of a QE hole which identifies the electrostatic center of thephotocathode. In addition, there is QE reduction across
the entire active area.

The same photocathode was used for all measurements. When QE had degraded sufficiently, the photocathode was moved to the load vacuum chamber, heated to ~ 580 C for up to 24 hours and reactivated. In total, the photocathode was heated and reactivated 5 times and a total charge of 1.3 kC was extracted for these measurements. In all cases, QE restored to initial values following the heat cycle.

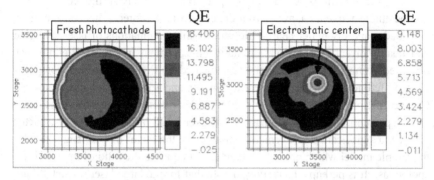

Figure 2. Two photocathode "QE scans", a) fresh photocathode following activation, b) after extracting 100 C. Plot axes are in arbitrary stepper motor units. The photocathode active area is 5 mm diameter.

3. Results

Charge lifetime results for the first 12 runs spanning the entire parameter space are shown in Figure 3a. For these runs, the best charge lifetimes (few hundred Coulombs) were obtained at 1.5mA for all laser spot sizes. At 10mA, charge lifetime was always very poor, just a few Coulombs. A very simple functional form was applied to each data set, a/I^b, where I is beam current, a is set equal to the charge lifetime value at 1.5 mA, and b is a parameter varied to provide the best fit to the data at higher currents. In part, the curve fitting merely helps the eye distinguish results for each spot size but it also represents an attempt to appreciate the mechanisms that contribute to ion formation and QE degradation. When $b=1$, a strict beam current dependence is implied which seems reasonable since at higher current, there are more electrons to ionize gas and more ions to damage the photocathode. For $b>1$, a current + vacuum dependence is implied, which also seems reasonable since more current produces more vacuum degradation from the dump and there are more stray electrons hitting the beampipe and liberating gas. For the results presented in Figure 3a, the curve fitting yields $b>1$ suggesting both current and vacuum dependence on lifetime. Charge lifetime results in Figure 3a are not particularly impressive and no indication of lifetime enhancement for larger laser spot sizes can be clearly identified. In retrospect, poor lifetime is not too surprising because a conscious attempt was made to run beam from near the center of the active area, which corresponds to the electrostatic center of the photocathode, the location where high energy ions are focused.

For subsequent runs, the laser spot was moved near the edge of the photocathode active area, away from the electrostatic center. This configuration provided markedly better charge lifetime for all combinations of laser spot size and beam current. Moreover, dramatic improvement in charge lifetime was observed using larger laser spots as indicated in Figure 3b which compares charge lifetime versus beam current for 342um and 1538um laser spot diameters. Charge lifetime between 1000 and 2000C was measured using the large laser spot, representing a lifetime enhancement over small laser spot data by a factor of ~ 10, a significant improvement although not the factor of ~ 20 suggested by the simple model which predicts enhancement equal to the ratio of the two laser spot areas. It is perhaps interesting to note that for the large laser diameter data, the simple curve fitting expression noted above provides fit parameter $b<1$, suggesting charge lifetime independent of current and vacuum, an interesting result that may or may not be relevant, but certainly requires more study.

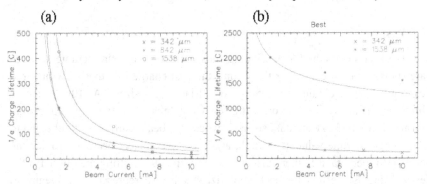

Figure 3. Charge lifetime results for different laser spot sizes and beam currents to 10 mA, with laser spot positioned: a) near the electrostatic center of photocathode, and b) near edge of photocathode active area.

4. Conclusions

Very high charge lifetimes were obtained at beam currents up to 10 mA using a 1538um diameter laser spot positioned away from the electrostatic center of the photocathode. The measured charge lifetime of 1500C at 10 mA is perhaps the highest value ever reported, higher in fact than CEBAF charge lifetime by a factor ~ 10 where maximum average beam current is ~ 200uA and laser spot size is ~ 500 um. There is clear evidence that lifetime improves using larger laser spot sizes, at least when the laser is positioned away from the electrostatic center, which should be very useful information for those building high current machines, however lifetime scaling did not match simple predictions. This submission represents the initial phase of a lengthy study of photocathode lifetime at high current, ultimately with high polarization photocathode material and rf-pulsed drive lasers.

Acknowledgments

This work is supported by U.S. DOE under contract DE-AC05-84ER401050.

References

1. V. Ptitsyn et al, "eRHIC, Future Electron-Ion Collider at BNL", Proceedings of the 2004 European Particle Accelerator Conference, Lucerne, Switzerland.
2. Y. Derbenev et al, "Electron-Ion Collider at CEBAF: New Insights and Conceptual Progress", Proceedings of the 2004 European Particle Accelerator Conference, Lucerne, Switzerland. L. Merminga et al, "ELIC: An Electron-Light Ion Collider based at CEBAF", Proceedings of the 2002 European Particle Accelerator Conference, Paris, France.
3. I. Bazarov et al, "Phase 1 Energy Recovery Linac at Cornell University", Proceedings of the 2002 European Particle Accelerator Conference, Paris, France.
4. M. Stutzman et al, "Status of JLab's Load Locked Polarized Electron Source", Proceedings of the 2002 Workshop on Polarized Electron Sources and Polarimeters, Middleton, MA, and J. Grames et al, "Ion Backbombardment of GaAs Photocathodes Inside DC High Voltage Electron Guns", Proceedings of the 2005 Particle Accelerator Conference, Knoxville, TN, 2005.
5. K. Aulenbacher et al, "Status of Polarized Source at MAMI", Proceedings of the 2002 Workshop on Polarized Electron Sources and Polarimeters, Middleton, MA.
6. http://www.eyeonscience.com

ILC @ SLAC R&D PROGRAM FOR A POLARIZED RF GUN[*]

J. E. CLENDENIN, A. BRACHMANN, D. H. DOWELL, E. L. GARWIN,
K. IOAKEIMIDI, R. E. KIRBY, T. MARUYAMA, R. A. MILLER, C. Y. PRESCOTT,
J. W. WANG

Stanford Linear Accelerator Center
Menlo Park, CA 94025, USA

J. W. LEWELLEN

Argonne National Laboratory
Argonne, IL 60439, USA

R. PREPOST

Department of Physics, University of Wisconsin
Madison, WI 53706, USA

Photocathode rf guns produce high-energy low-emittance electron beams. DC guns utilizing GaAs photocathodes have proven successful for generating polarized electron beams for accelerators, but they require rf bunching systems that significantly increase the transverse emittance of the beam. With higher extraction field and beam energy, rf guns can support higher current densities at the cathode. The source laser system can then be used to generate the high peak current, relatively low duty-factor micropulses required by the ILC without the need for post-extraction rf bunching. The net result is that the injection system for a polarized rf gun can be identical to that for an unpolarized rf gun. However, there is some uncertainty as to the survivability of an activated GaAs cathode in the environment of an operating rf gun. Consequently, before attempting to design a polarized rf gun for the ILC, SLAC plans to develop an rf test gun to demonstrate the rf operating conditions suitable for an activated GaAs cathode.

1. Introduction

The ultimate goal of the R&D program is to develop an L-band rf gun system for polarized electrons that will meet the ILC beam operational requirements. Unpolarized rf photoinjectors have already been demonstrated to meet these requirements [1], but because of the uncertainty in the viability of an activated GaAs-type photocathode in the environment of an operating rf gun, the ILC

---

[*] Work supported by Department of Energy contracts DE-AC02-76SF00515 (SLAC), W-31-109-ENG-38 (ANL) and DE-AC02-76ER00881 (UW).

baseline configuration presently specifies a dc gun for generation of polarized electrons. The advantages of an rf over a dc gun include:

- For a beam that consists of a train of closely-spaced pulses (as required by the ILC), the source laser system can be used to generate the high peak current required for each micropulse without the need for post-extraction rf chopping or bunching;
- Lower beam emittance—both transverse and longitudinal—which will improve the operational reliability and efficiency of the injector; and
- Higher quantum yield (QY) with a higher threshold for the surface charge limit.

2. Elements of the R&D Program

2.1. *Improved Vacuum*

Operating pressures at the cathode of 10^{-11} Torr or better are essential. To achieve this level, the conductance between the cathode and pump must be significantly improved and the outgassing rate of the structure must be decreased. The conductance for pumping on any rf structure can be improved by using z slots [2] or multiple small holes (sieve) [3] in the outer cylinder. In addition, the conductance within the rf structure itself depends on the design. Two interesting possibilities are the plane-wave-transformer (PWT) and the higher-order-mode (HOM) designs. The HOM design [4] is particularly inviting because it contains no internal structures, which not only optimizes conductance, but also minimizes internal joints and simplifies cooling. By choosing an appropriately tapered radius for the HOM structure, it's electric field on axis, $E_z(z,r=0)$, can be made equivalent to that of the standard BNL-type design. A tapered 0.75-wavelength L-band HOM cavity is compared to a standard 0.715-wavelength structure in Figure

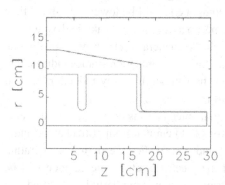

Figure 1. Cross section of a tapered L-band HOM TM011 rf gun structure (larger radius) superimposed over a standard BNL-type TM010 gun structure.

1. An HOM gun with slots or sieve would have an overall conductance about 20 times higher than a standard gun.

The typical gas load of an operating rf gun corresponds to an outgassing rate of 10^{-11} Torr-l/s cm^{-2}, which includes the effect of virtual leaks from grain boundaries and structural joints. A reduction in this rate of at least an order of magnitude is required. The type of material used may be important. Various types of Class 1 OFE Cu and related alloys will be examined. Assembly and cleaning techniques will also be evaluated.

Improved conductance and a lower outgassing rate, combined with sufficient pumping speed—provided at these low pressures by NEG pumps—should result in the desired pressure at the cathode.

2.2. Reduction of Field Emitted Electrons

When a GaAs cathode activated to negative electron affinity (NEA) was tested in a ½-cell S-band rf gun at BINP, the QY lifetime of the cathode was measured in terms of only a few rf pulses [5]. This rapid deterioration was attributed to back bombardment of the cathode by electrons, which of course does not occur for a dc gun. Since all the photoelectrons are expected to exit the gun, the principal problem will be with field emitted electrons. Field emission is most likely to originate at the cathode plug, the iris, and the rf input coupler. The coupler source can be mitigated by using z-coupling or eliminated entirely using axial coupling. The iris source can be reduced by elliptical shaping. Field emission from the cathode plug is a problem that will require special study. Of course the peak rf fields for all these components will be lower for L-band than for S-band. In addition, the gun can be operated at the peak field value that minimizes damage to the cathode from back accelerated electrons. As with a dc gun, ions can also damage the cathode. However simulation studies indicate that few if any of the ions will actually hit the cathode [6]. Any reduction in field emission will also reduce the number of ions present.

To understand better the potential mechanisms for cathode damage, simulations will be used to track electrons field emitted from critical areas under a variety of conditions (peak rf field and phase). For the optimum operating conditions, the energy distribution of the electrons hitting the cathode will be determined. Although the number distribution cannot similarly be specified, the energy information will be useful to guide very controlled cathode damage studies using SLAC's surface analysis apparatus [7].

Simulations will also be used to determine the operating peak rf field and phase that will minimize cathode surface damage.

3. Testing

Once the elements of the program described in section 2 above have matured, a test gun will be constructed paying particular attention to the elements that affect the cathode performance, but to save time and cost, minimizing efforts on issues that are already understood. A high quality cathode preparation chamber and load-lock will be integrated into the gun design.

The test gun will be rf processed with a dummy cathode in place. The L-band rf station now under construction at SLAC will provide the required rf power. Following successful processing, the QY lifetime of an activated cathode placed in the gun will be measured without rf. Lifetimes comparable to those achieved with dc guns should be demonstrated. This same system will be studied with rf on to determine what the limits are on the QY lifetime of the activated cathode relative to no rf and what the sources are of any deteriorated performance.

4. Conclusion

The R&D program at SLAC for a polarized rf gun is designed to demonstrate the viability of such a gun system for the ILC injector. The major known problems will each be studied and optimum solutions incorporated into the design of a test gun. A demonstration of reasonable QY lifetime with rf on will provide the technical justification for the construction of a polarized rf gun that will meet all the operational requirements—in addition to polarization—of the ILC.

References

1. J. Baehr et al., *TESLA Note 2003-33* (2003).
2. R. Sheffield et al., *PAC93*, p. 2970.
3. D. Yu et al., *PAC2003*, p. 2129.
4. J. W. Lewellen, *Phys. Rev. ST-AB* **4**, 040101 (2001); J. W. Lewellen, *LINAC2002*, p. 671.
5. A. Aleksandrov et al., *EPAC98*, p. 1450.
6. J. W. Lewellen, *Phys. Rev. ST-AB* **5**, 020101 (2002); R. P. Fliller III et al., *PAC2005*, p. 2708.
7. D. T. Palmer, R. E. Kirby, F. K. King, *SLAC-PUB-11355* (2005).

HIGH FIELD GRADIENT POLARIZED ELECTRON GUN FOR ILC

M. YAMAMOTO, N. YAMAMOTO, T. NAKANISHI, S. OKUMI,
M. KUWAHARA, K. YASUI, T. MORINO, R. SAKAI, K. TAMAGAKI

Department of Physics, Graduate School of Science, Nagoya University,
Nagoya, 464-8602, Japan
E-mail: yamamoto@spin.phys.nagoya-u.ac.jp

F. FURUTA, M. KURIKI, H. MATSUMOTO, M. YOSHIOKA

High Energy Accelerator Research Organization (KEK),
Tsukuba, 305-0801, Japan

A 200-keV gun has been developed for generation of bunch charge of $\geq$ 3.2 nC, bunch length of $\leq$ 2 ns multi-bunch polarized electron beam that is required for International Linear Collider. In this paper, the beam simulations using General Particle Tracer (GPT) code for such space-charge dominated regions, and a fabrication method of electrode for higher electric field ($\geq$ 3 MV/m) at the photocathode surface by using a titanium anode and a molybdenum cathode are described.

1. Introduction

International Linear Collider (ILC) requires high bunch charge ($\geq$3.2 nC/bunch), multi-bunch structure ($\sim$2 ns bunch length, 337 ns separation), high spin polarization ($\geq$80%) and low-emittance beam at the source[1].

The detail of bunch parameters will be optimized by a multi-bunch laser system, charge limit restrictions (space charge limit, NEA surface charge limit), beam transmission efficiency and emittance at downstream of a pre-buncher/buncher section.

Higher gun voltage and electric field have advantages of a generation such a high-intensity and low-emittance beam. However, it becomes more difficult to suppress a field emission current and to maintain long lifetime of a NEA surface. Therefore, it is indispensable to develop a fabrication technology of electrode that can suppress the field emission ($\leq$ 10 nA) in high gun voltage and high field gradient conditions.

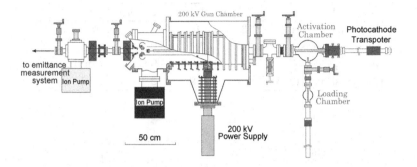

Figure 1. Schematic view of the 200-keV polarized electron source

2. Emittance Simulation

The 200-keV polarized electron gun[2] and a strained GaAs/GaAsP superlattice photocathode[3,4] have been developed for applications of ILC. A schematic view of the 200-keV gun is shown in Figure 1.

Beam simulations which the gun operated 200 kV were done by using General Particle Tracer (GPT) code[5]. Field maps along the beam line (static electric field of the electrodes and magnetic field of a solenoid) were calculated using POISSON code[6].

The transverse and longitudinal laser profiles were assumed to be gaussian distribution. These widths were defined within a 95 % area of the total laser energy.

The laser spot size of $\phi 18mm$ was chosen to reduce the surface-charge-limit effect peculiar for NEA surface.

The laser pulse duration was assumed to be between 0.7 ns and 2.0 ns. The solenoid was positioned 13 cm downstream from the photocathode. All calculations were done using optimized magnetic field of the solenoid so that the normalized rms emittance was minimized at 50 cm downstream from the photocathode. Results presented here are assuming to 500 macro-particles in one bunch by using a cylindrically symmetric 2D space-charge approximation that corresponds the number of macro-particles ($\geq 10^4$) 3D space-charge calculation.

The result shows that the 200 keV gun can deliver 0.7ns bunch beam with more than 99% beam transmission efficiency and the normalized rms emittance is less than $15\pi mm \cdot mrad$ by $\sim$5 nC bunch charge.

Laser pulse duration	0.7 ns	1.5 ns	2.0 ns	Units
Bunch length (rms)	43	86	115	[mm]
Average radius	3.9	3.4	3.2	[mm]
Norm. $\epsilon_{r.rms}$	12.9	10.0	9.0	[π.mm.mrad]
Energy spread (rms)	6.4	2.9	2.4	[keV]

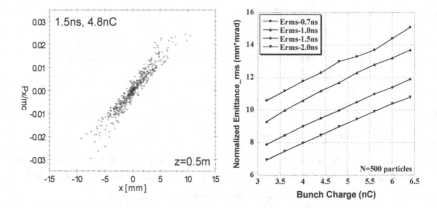

Figure 2. GPT simulations of distribution in horizontal phase space (left), the rms emittance as a function of the bunch charge between 3.2 nC and 6.4 nC for the bunch duration 0.7, 1.0, 1.5, 2.0 ns, respectively.

3. New Materials for the Gun Electrode

New material electrodes made of molybdenum and titanium have been developed to suppress the field emission dark current in higher voltage for a restriction of increase the emittance by the space-charge force.

It is well-known that positive ions and residual molecules produced by collisions of dark current from the electrode degrade NEA surface. Therefore, the reduction of the dark current is a key technology for development of a high field gradient polarized electron gun.

The field gradient of the photocathode surface is 3 MV/m and the electrode surface maximum is 7.8 MV/m for 200 kV operation. In order to applied higher voltage ($\geq$200kV) or higher field gradient ($\geq$8MV/m) at the electrode, a new electrode material was required to avoid such a dark current problem. Using a test apparatus at KEK, it was demonstrated that a combination of a molybdenum cathode and a titanium anode is the best to suppress the field emission dark current[7]. A large molybdenum cathode

(~160 mm diameter) is employed to fabricate a hot-spinning method using a 2 mm thickness molybdenum plate (purity:≥ 99.96%).

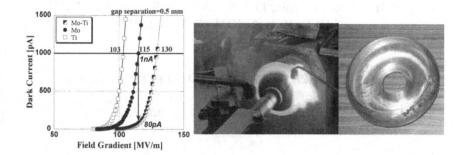

Figure 3. A comparison of field emission dark current between Ti-Ti,Mo-Mo and Mo-Ti electrode (cathode-anode) (left), pictures of fabricated by a hot-spinning procedure and the finished figure.

4. Conclusion

Typical ILC beam simulations for the 200-keV gun by using GPT code were done. Improvement of the 200-keV gun voltage and increase the field gradient by using the molybdenum cathode and the titanium anode is in progress. This technology will be useful to build in a further ultra-low emittance ($\leq 1.0\pi \cdot mm \cdot mrad$) electron source.

References

1. Strawman Baseline Configuration Document "electron source parameter".
2. K. Wada, et.al., *AIP Conf.***675**, 1063 (2002).
3. O. Watanabe, T. Nishitani, et.al., *AIP Conf.***570**, 1024 (2000).
4. T. Nishitani, T. Nakanishi, et.al., *J. Appl. Phys.***97**, 094907 (2005).
5. General Particle Tracer (GPT), release 2.52, Pulsar Physics.
6. J. Billen and L. Young, Los Alamos Laboratory Technical Report No. LA-UR-96-1834, 2000.
7. F. Furuta, T. Nakanishi et.al., *Nucl. Instrum. Meth.***A 538**, 33 (2005).

POLARIZED ELECTRON SOURCES FOR FUTURE ELECTRON ION COLLIDERS *

M. FARKHONDEH, W. FRANKLIN, AND E. TSENTALOVICH

MIT-Bates Linear Accelerator Center,
P.O.Box 846, Middleton, MA 01949, USA
E-mail: manouch@mit.edu

ILAN BEN-ZVI AND V. LITVINENKO

Brookhaven National Laboratory,
Upton, NY, 11973, USA

Studies are underway in the US to design an electron-ion collider (EIC) with high luminosity (10^{-33} cm^{-2}s^{-1}) and high center-of-mass energy centered around the existing nuclear physics accelerators at either BNL or Jefferson Lab. EIC will be optimized for studying the formation and structure of hadrons in terms of their quark and gluon constituents. For the RHIC-based collider, eRHIC, two options are under consideration: a more mature option based on a ring-ring concept with e-p and e-A collisions, and a more conceptual design with a linac-ring architecture. Ions from the existing RHIC hadron ring will collide with 5-10 GeV electrons, either from a new electron storage ring or a very high current cw linac. In both options, high current electron bunches in the polarized injector must be precisely synchronized with proton or ion bunches in the RHIC ring. The stacking option in the ring-ring design considerably reduces the required bunch charge from the polarized source. The polarized source requirements for the eRHIC linac-ring design are very demanding, requiring sources capable of producing highly polarized cw currents of 200-300 mA. In this paper, we present, for the two options of eRHIC, the polarized source requirements and the design parameters of the laser systems considered for these sources. We also present the current R& D issues for the eRHIC polarized sources.

1. Introduction

Lepton probes with high luminosities are very suitable for answering questions about the structure of hadrons in terms of their quark and gluon

---

*This work is supported by the U.S. Department of Energy under a Cooperative Agreement # BEFC294ER40818.

constituents, and the evolution of quarks and gluons into hadrons. An Electron-Ion Collider (EIC) can directly probe the quark gluon distribution in nucleons over a wide range of x, the fraction of proton momentum carried by the struck quark and gluon. Such a facility will need to have high center-of-mass energy, high luminosity for precision, polarized leptons and polarized nucleons, a complete range of hadron beams and optimum detectors. Substantial international interest in high luminosity (10^{33} cm^{-2}s^{-1}) polarized lepton-ion colliders over the past decade resulted in several international workshops in the US and in Europe. An important aspect of EIC is the polarization of both lepton and ion beams. One of the major initiatives for EIC in the US is based on the existing RHIC accelerator at BNL. There are two design considerations for EIC using RHIC:

(1) **eRHIC Ring-Ring design:** An EIC based on a new electron ring and the existing RHIC facility.
(2) **eRHIC Linac-Ring Concept:** An EIC based on a high current energy recovery linac (ERL) and the existing RHIC facility.

In recent years at Jefferson-Lab, a concept for an Electron Light-Ion Collider emerged[2] that is based on a 3–7 GeV ERL electron linac, a new electron circulator ring and a new light ion ring. In 2004, physicists from several institutions, including BNL and MIT-Bates, produced a comprehensive zeroth order design report (ZDR)[1] for eRHIC. In the next two sections, we present polarized source requirements for the two eRHIC concepts. Also, a preliminary polarized source design developed at MIT will be presented for the eRHIC injector.

2. Polarized Source for eRHIC

2.1. *Ring-Ring Design*

The present main design for eRHIC calls for a 0.5 A electron storage ring with full-energy injection from a linac with a polarized electron source. Although the electron ring will be self-polarizing, the polarization build-up time for 5 GeV beams will be several hours, meaning that the ring must be injected with a highly polarized beam. Furthermore, use of a highly polarized source will allow operation of the storage ring in top-off mode, permitting the electron beam intensity to remain high at all times. For a CW storage ring, the achievement of 0.5 A of highly polarized electrons would represent a modest technical requirement based on present state-of-the-art polarized source technology. However, because eRHIC is a collider, synchronized bunches of electrons must precisely match the time structure of the hadron bunches in the RHIC ring. This presents a great challenge to the injector configuration and the polarized source design. The polarized source must address two primary challenges: the time and bunch structure of the ring-ring collider, and the necessary laser power required to achieve the charge per bunch for the stated luminosity for the collider. A detailed evaluation of the eRHIC luminosity (10^{33}cm^{-2}s^{-1}) design value shows that peak currents of at least 20 mA from the source are required. The corresponding charge per bunch is 1.3 pC for bunches 70 picoseconds long produced at 28 MHz synchronous with the collider ring. This is a challenging technical requirement for a photoinjector. The specifications of the eRHIC ring-ring option are summarized in Table 1.

Table 1. Beam specification for the eRHIC ring-ring polarized electron source design.

	Quantity	Value	Unit
Collider ring	Stored current	480	mA
	Frequency	≈ 28	MHz
	Ring circumference	4.3	μs
	Number of bunches	120	
	Charge per macro bunch	20	nC
(stacking)	pulse train rep. rate	25	Hz
	Duration	10	minutes
	Total pulse train	15,000	
Photoinjector	Bunch duration	70	ps
	Bunch charge	1.3	pC
	Peak current	20	mA

Photoemission in the eRHIC polarized electron injector would be produced by illuminating high polarization GaAs-based photocathodes with circularly polarized laser light at 780–830 nm. For this range of wavelengths, a laser peak power of at least 50 W would be needed for the beam specifications given in Table 1, assuming a quantum efficiency (QE) of order 5×10^{-3}. Currently, designs based on two different types of laser systems are being considered[3]. These two options differ in the time structure of the photoemission drive laser systems and in the electron beam line for bunching and chopping functions. The first option is based on a mode-locked diode laser [4] capable of producing laser bunches synchronized with pulses in the storage ring. A schematic diagram of the first option based on a mode-locked laser system is shown in Figure 1.

The second option relies on a high-powered DC fiber-coupled diode array

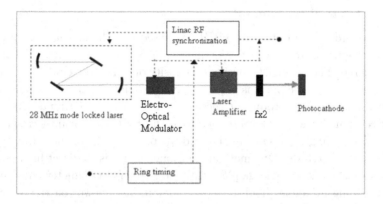

Figure 1. Schematic diagram of the mode-locked laser option for the eRHIC polarized injector.

laser system similar to one employed for the MIT-Bates polarized source[5]. In this case, bunching and chopping elements in the linac injector are used to produce bunches synchronized with the collider ring. This option employs a commercial high power diode laser system. A schematic view of this option is shown in Figure 2. The laser produces DC or pulsed radiation with no microscopic structure. The challenge of this option is to ascertain the degree to which the complex bunching and chopping of the electron beam at multiple frequencies is possible.

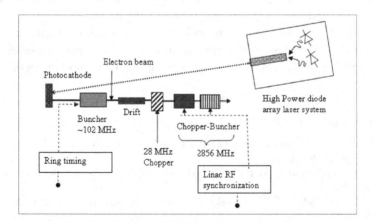

Figure 2. Schematic diagram of the laser and the electron beam layout for the DC fiber-coupled diode laser option for the eRHIC injector (Option Two).

2.2. Linac-Ring Concept

The second design option for eRHIC is based on a linac-ring concept. In this design, a 5–10 GeV superconducting energy recovery linac (ERL)[1] is considered. For the eRHIC design luminosity of order 10^{33} cm^{-2}s^{-1} and an average current of 400 mA is required from the linac. Therefore, a polarized injector capable of producing these very high average currents with the correct time structure synchronous with the RHIC ion beam is required. This demanding requirement is beyond the present state of the art by three orders of magnitude. One method envisioned so far is to direct high power lasers onto a relatively large photocathode 7–10 cm in diameter. Assuming that scaling principles hold in photoemission, laser power on the order of 1 kW is required for this injector. One concept for such a laser system, described in Appendix A of the ZDR[1] document, is the use of the output of a dedicated low energy ERL-FEL at 780–830 nm range as the drive laser for this collider. To produce 100% circularly polarized light, an FEL based on a helical wiggler must be used. Photocathode surface charge limit effects needs to be studied for these high power laser densities. Also of concern is the heat dissipation in the photocathode by the drive laser. A 3 cm^2 laser spot with 1 kW power on a 1 mm thick GaAs based photocathode with heat conductivity k=0.75 w/cm.°C can create a temperature rise of order 40 °C. This is an adverse effect for the delicately balanced Cs activated NEA surface of the photocathode and the UHV condition of the gun chamber. Active cooling with liquid or cold gas is essential to keep the photocathode and the molybdenum cathode stalk from heating up and destroyng the QE.

Areas of R&D for this source include a) design and construction of an actively cooled large area photocathode assembly and b) photoemission and charge limit effect studies with this system using high power lasers of order kW.

References

1. Zeroth order Design Report (ZDR), http://www.agsrhichome.bnl.gov/eRHIC/eRHIC_ZDR/ZDR_start.pdf
2. L. Merminga and Y. Derbenev, Proc. European Particle Accelerator Conference (2002).
3. M. Farkhondeh, 2nd EIC Accelerator Workshop, Jefferson Lab, March 15-17, 2004.
4. Time-Bandwidth mode-locked laser for the G0 experiment at Jefferson Lab.
5. Spectra-Physics Opto Power diode laser model OPC-DO60-mmm-FC. See also E. Tsentalovich, AIP Conf. Proc. 675 (2002) 1019.

COMPARISON OF ALINGAAS/GAAS SUPERLATTICE PHOTOCATHODES HAVING LOW CONDUCTION BAND OFFSET[*]

K. IOAKEIMIDI, T. MARUYAMA, J. E. CLENDENIN, A. BRACHMANN, E. L. GARWIN, R. E. KIRBY, C. Y. PRESCOTT, D. VASILYEV
Stanford Linear Accelerator Center
Menlo Park, CA 94025, USA

Y. A. MAMAEV, L. G. GERCHIKOV, A.V. SUBASHIEV, Y. P. YASHIN
Saint-Petersburg State Polytechnic University, Politechnicheskaya 29, Saint-Petersburg, Russia, 195251

R. PREPOST
Department of Physics, University of Wisconsin
Madison, WI 53706, USA

The main advantage of superlattice (SL) structures as spin polarized electron emitters is the ability to provide a large splitting between the heavy hole (HH) and light hole (LH) valence bands (VB) over a large active thickness compared to single strained layers. Two important depolarization mechanisms in these structures are the scattering effects during the transit of the electrons in the active region and the depolarization that takes place in the band bending region (BBR) near the surface. In this paper, we systematically study the effects of the electron mobility and transit time by using an InAlGaAs/GaAs SL with a flat conduction band (CB). Initial results by the SPTU-SLAC collaboration using such structures grown by the Ioffe Institute showed polarization and quantum yield (QE) of 92% and 0.2% respectively. We report measurements using similar structures grown by SVT Associates. The results (polarization up to 90%) are also compared with simulations.

1. Introduction

High polarization electron sources are an important part of the International Linear Collider effort at SLAC. In previous work, polarization on the order of 90% was achieved with the GaAs/GaAsP SL [1], [2].
The main spin depolarization mechanisms in these structures are:

---

[*] Work supported by Department of Energy contracts DE-AC02-76SF00515 (SLAC) and DE-AC02-76ER00881 (UW), RFBR under grant 04-02-16038 and NATO under grant PST.CLG.979966

1. Interband absorption smearing δ due to bandedge fluctuations;
2. Hole scattering between the HH and LH states that causes a broadening γ of the LH band;
3. Spin precession due to an effective magnetic field generated by the lack of crystal inversion symmetry and spin orbit coupling;
4. Electron-hole scattering (negligible comparing to 3);
5. Less polarization selectivity in the BBR;
6. Scattering and trapping of electrons in the BBR.

The first two mechanisms are related to the HH-LH splitting for supporting the spin selection rules. A systematic study on the GaAs/GaAsP structure [2] showed that after a certain splitting level, no increase of polarization could be obtained. Mechanisms 5 and 6 are related to the effects of the BBR and will be independently studied in the future. Mechanisms 3 and 4 are material related and they take place during the transport of electrons in the photocathode active region. In the GaAs/GaAsP SL, the electrons tunnel through high barriers in order to reach the cathode surface. In order to lower the barriers in the CB without lowering the barriers for the holes in the VB (to preserve the HH-LH splitting), a quaternary alloy InGaAlAs/GaAs SL was designed and tested at St. Petersburg University, and polarization as high as 91% was achieved with an optimized structure [3]. The measurements were repeated at SLAC on samples grown by SVT Associates. The results are presented in this paper and they are compared with simulations.

2. Design of Flat Conduction Band SL Structures

The model for the emitted electron polarization [4] indicates that polarization is inversely proportional to the electron transit time in the active region. Motivated by this concept, flat CB structures based on $In_xAl_yGa_{1-x-y}As/GaAs$ strained barrier SL were designed. The x (In) percentage lowers the bandgap, controls the CB offset ΔE_C and induces compressive strain in the barriers in order to achieve the desirable HH-LH splitting. The y (Al) percentage controls the size of the SL bandgap and preserves high barriers for the holes in the VB. The goal is to design a structure with as flat a CB as possible, while maintaining a substantial (>30meV) VB splitting. The CB gets anomalously flat for x=1.1y. The VB splitting is determined by the induced strain in the barriers controlled by the Indium percentage and the quantum confinement of the wells controlled by the barrier/well sizes. For the $Al_{0.21}In_{0.20}Ga_{0.59}As/GaAs$ SL with 1.5nm wells and 4nm barriers the HH-LH splitting is >50meV.

3. Experimental Results

The parameters of the measured samples are shown in Table 1. The first 3 samples are grown by SVT Associates and the last 3 by the Ioffe Institute. All samples have 1.5nm quantum well width, 4nm barrier width, 18 periods, and $4\times10^{17}cm^{-3}$ Be doping. The BBR thickness is 6nm. For comparison the GaAsP/GaAs SL has 89meV LH-HH splitting, ΔE_C=97meV.

Table 1. Parameters of measured samples

Sample	In%	Al%	SL BG	BBR dop	LH-HH	ΔE_C	Polarization %
5506	17	18	1.449 eV	1e19cm^{-3}	52meV	19meV	82-85
5501	20	21	1.454 eV	1e19cm^{-3}	70meV	19meV	84-90
5503	23	25	1.469 eV	1e19cm^{-3}	68meV	10meV	75-82
5-777	20	23	1.471 eV	1e19cm^{-3}	60meV	3meV	91
6-329	20	22	1.463 eV	7e18cm^{-3}	61meV	11meV	76-78
6-410	28	35	1.542 eV	7e18cm^{-3}	90meV	23meV	75-82

Experimental results (dots, multiple measurements) along with the simulations (solid lines) for samples #5501, #5-777 are shown in Figures 1,2.

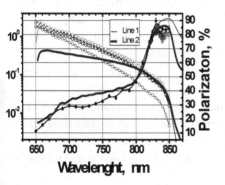

Figure 1. Sample #5501.

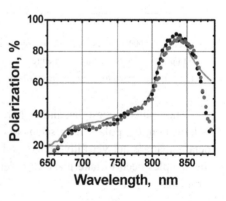

Figure 2. Sample #5-777.

4. Discussion and Conclusions

The measured peak polarization is shown in Table 1. By comparing the data with simulation results for samples # 5501 (line 1, Figure 1), 5503 and 5506, we observe a blue shift for the experimental peak. Although longer wavelength photons in general provide more spin selectivity, they also photogenerate electrons primarily in the BBR where there is less polarization selectivity. Also, the electrons photogenerated in the SL structure by longer wavelengths thermalize faster and get trapped more easily in the BBR where they depolarize. Simulation results match the polarization peak height when depolarization is considered to take place in the BBR (line2, figure 1).

All the Ioffe samples were protected by As caps. The highest polarization (91%) was measured when sample #5-777 was heat cleaned at 450°C, while the peak polarization dropped to 85% after heat cleaning at 540°C. The SVT samples where activated after being heat cleaned at 540°C. The effect of the heat cleaning temperature on the polarization suggests that there is a surface factor that contributes to the depolarization. One possible explanation is that the SVT samples have a broader BBR than sample #5-777 due to higher heat cleaning. Samples #6-329 and #6-410 have lower doping at the surface layer and thus, broader BBR than #5-777 and they don't achieve high polarization

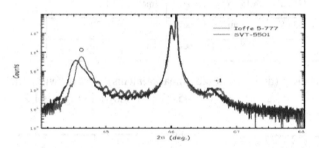

Figure 3. (004) X-Ray analysis of samples 5501 and 5-777. Sample 5-777 has a "cleaner" structure and slightly smaller period (5.01nm) compared to sample 5501 (5.10nm).

As shown in the (004) x-ray results of Figure 3, the In concentration is slightly higher in the #5501 sample. The deformed SL structure of the SVTA samples can contribute to the lower polarization due to absorption smearing.

The results suggest that although the flat CB samples are promising for high polarization, the polarization seems to depend on surface effects and structural details that are not yet fully understood. Further studies and SIMS analysis of the samples need to take place in order to draw final conclusions about these structures.

References

1. T. Nishitani, et al., *J. Appl. Phys.*, **97**, 094907 (2005).
2. T. Maruyama et al., *Appl. Phys. Lett.*, **85**, 2640 (2004).
3. Yu. Mamaev et al., *SPIN* 2004, p. 913, SLAC-PUB-10891
4. A.V. Subashiev et al., *SLAC-PUB*-7995, PPRC-TN-98-6, Nov 1998.

GENERATION OF SPIN POLARIZED ELECTRONS BY FIELD EMISSION

M. KUWAHARA, T. NAKANISHI, S. OKUMI, M. YAMAMOTO, M. MIYAMOTO,
N. YAMAMOTO, K. YASUI, T. MORINO, R. SAKAI, K TAMAGAKI
Graduate School of Science, Nagoya University
Nagoya, 464-8602, Japan.

K. YAMAGUCHI
Department of Electronic Engineering, The University of Electro-Communications
Tokyo 182-8585, Japan.

A pyramidal shaped GaAs (tip-GaAs) photocathode for a polarized electron source (PES) was developed to improve beam brightness and negative electron affinity (NEA) lifetime by using field emission. The emission mechanism also enables the photocathode to extract electrons from the positive electron affinity (PEA) surface, and relax the NEA lifetime problem. I-V characteristics of electrons extracted from tip-GaAs shows that the electron beam was extracted by field emission mechanism, because a linear dependence was obtained in Fowler-Nordheim (F-N) plot. Furthermore, a tip-GaAs cathode has succeeded in generation of spin polarized electron beam. The polarization degree of tip-GaAs is about 34% at excitation photon energy of 1.63eV which is no less than that obtained by an NEA-GaAs cathode.

1. Introduction

Strained-layer superlattice structures have been presenting the most promising performance as a photocathode for the polarized electron source (PES). In our experiments, the GaAs-GaAsP photocathode achieved maximum polarization of 92±6% with quantum efficiency of 0.5%, while the InGaAs-AlGaAs photocathode provided higher quantum efficiency (0.7%) with lower polarization (77±5%). Criteria for achieving high spin polarization and high quantum efficiency using superlattice photocathodes were clarified by employing the spin-resolved quantum efficiency spectra. [1-3]

However, it seems that major problems remained for the PES R&D are to improve (1) beam emittance and (2) NEA lifetime under gun operations for high peak current and high average current, respectively. In order to overcome these problems simultaneously, we started a development of a new type photocathode using field emission mechanism. First, we tried to use a pyramidal shape GaAs (tip-GaAs). Using the tip-GaAs, electrons can be emitted from a small area at the top of pyramid, and thus the beam emittance is expected to be small. This emission mechanism also enables to extract electrons from the poor NEA or small PEA surface into vacuum, and it helps to relax the NEA lifetime problem.

2. Experimental Procedure

The pyramidal shaped GaAs was fabricated from the Zn-doped GaAs (100) substrate by anisotropic wet etching using H_3PO_4 solutions.[4] For GaAs etching, resist-mask patterns were prepared on the GaAs substrates by photolithography. The resist pattern was square, and length of side was 10μm. The edge of the square mask was aligned along the <010> direction. In order to sharpen the tip-radius, the GaAs etching was carried out in a $10H_3PO_4:H_2O_2:H_2O$ solution at −1 °C. As shown in Figure 1, a tip radius was about 25nm, and a distance between tip to tip was 200 μm. After this process, tip-GaAs was rinsed by a HCl-isopropanol treatment for removing gallium and arsenic oxides from the surface.[5]

In the experiment, we used a 20 kV DC gun and a 70 keV PES system.[2] The 20 kV DC gun for which the gap separation of electrodes was variable. The 20 kV DC gun could apply a high-gradient dc field of 3.8 MV/m to the photocathode surface under UHV (10^{-11} Torr) condition.

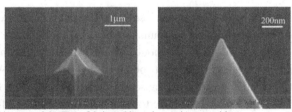

Figure 1. SEM images of tip-GaAs fabricated by anisotropic wet etching

3. Experimental Results and Discussions

The electrical characteristics was measured by 20 kV DC gun with illuminating laser light. Figure 2(a) shows the F-N plot using observed I-V characteristics of tip-GaAs. The data was roughly fitted by a straight line, which is inversely proportional to 1/E. This behavior suggests that the excited electron was extracted into vacuum by tunneling effect. Figure 2(b) shows the QE as a function of excitation photon energy under a high gradient field of 3.4MV/m. The solid line in figure 2(b) is a fitting curve using a calculation of tunneling yield based on WKB approximation. The QE is rising quickly at 1.6 eV. Such a behavior is not observed for the bulk-GaAs with NEA surface. Figure 7 shows ESP and QE of tip-GaAs under applying high gradient field and illuminating circularly polarized laser light together with the ESP of NEA-GaAs. The ESP of tip-GaAs has bumpy structures between 20 % and 40 %, but it has higher values than bulk-GaAs in wavelength region below 760 nm. This wavelength corresponds to the rising edge of QE at 1.6eV in photon energy range. These experimental results indicate that the spin polarized electrons can be extracted from the conduction band into vacuum through the tip of GaAs without serious depolarization. In order to understand these phenomena, we considered the depolarization mechanism in GaAs crystal and extraction mechanism. While drifting to the surface, partial excited electrons are scattered by holes, phonons, and so on. When the electrons excited above the bottom of the conduction band, the energy dispersion will be widen, and the low energy end of this wide dispersion has a lower polarization compared to the initial polarization. Since the tunneling yield of surface increases at a rapid rate for electron energy, electrons transmitting into a vacuum constitute the majority of high-energy end of wide dispersion. Therefore it would appear that the low-polarization portion is cut off, and high-polarization portion can only extract into vacuum.

4. Conclusion

We have demonstrated that spin polarized electrons can be extracted from tip-GaAs by using a field emission mechanism under circular light irradiation. The polarization degree of tip-GaAs is about 34% at excitation photon energy of 1.63eV which is no less than that obtained by an NEA-GaAs cathode. Furthermore, distinctive phenomena caused by field emission were observed in the ESP and QE spectrum. It is hoped that this photocathode will be widely applicable to accelerators analytical instruments.

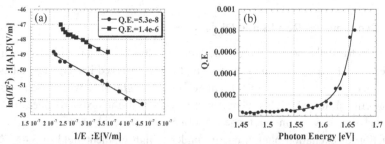

Figure 2. (a) F-N plot of photo-electron extracted from yip-GaAs, (b) QE of tip-GaAs under applying high gradient field as a function of excitation photon energy.

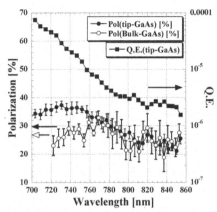

Figure 3. ESP and QE of tip-GaAs as a function of wavelength. ESP of NEA-GaAs was also plotted for comparison.

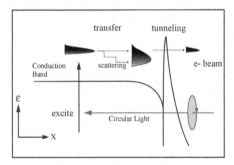

Figure 4. Schematic one-dimensional potential model of the field emission in a p-doped semiconductor. The magnitude of polarization is indicated by the light and shade region under the energy distribution curve, where darker shades correspond to higher polarization.

References

1. T. Nishitani et al., *J. Appl. Phys.* **97**, 094907 (2005).
2. K. Togawa, et al., *Nucl. Instr. Meth.* **A455**, 109 (2000).
3. K. Togawa, et al., *Nucl. Instr. Meth.* **A414**, 431 (1998).
4. K. Yamaguchi, et al., *J. Electrochem. Soc.* **Vol. 143, No. 8**, 2616 (1996).
5. O. E. Tereschchenko, et al., *Appl.Surf. Sci.* **V142,** 75 (1999).

Polarization in RI beam experiments

POLARIZED PROTON SOLID TARGET FOR RI BEAM EXPERIMENTS

T. WAKUI

Cyclotron and Radioisotope Center, Tohoku University,
6-3 Aoba, Aramaki, Aoba-ku,
Sendai, Miyagi 980-8578, Japan
E-mail: wakui@cyric.tohoku.ac.jp

A polarized solid proton target has been developed for experiments with radioactive isotope beams. The polarized target was successfully applied to scattering experiments with unstable ^{6}He beams. Average value of proton polarization in the scattering experiment was 13.8(39)%.

1. Introduction

A polarized solid proton target using a crystal of naphthalene doped with pentacene has been developed at the Center for Nuclear Study (CNS), the University of Tokyo.[1] The target is unique in that protons can be polarized at a low magnetic field of 0.1 T and a higher temperature of 100 K by combining methods called "microwave-induced optical nuclear polarization" and "integrated solid effect".[2] This feature enables the target to be used in a scattering experiment with a radioactive isotope (RI) beam under the inverse kinematic condition.[3]

The polarized proton target has been used in scattering experiments with unstable ^{6}He beams at the RIKEN projectile fragment separator (RIPS) in July 2003 and in July 2005. The experiment performed in July 2003 was the first scattering experiment by a combination of an RI beam and a polarized solid proton target.[4] In both experiments, the vector analyzing power for the elastic scattering of polarized proton and ^{6}He at the energy of 71 MeV/u was measured for θ_{cm}=40° − 78°. This article describe the polarized proton target system and its performance during the scattering experiment in 2005.

2. Target system

A target crystal is fabricated by a vertical Bridgman method after refining the naphthalene powder by the zone-melting method. The fabricated crystal is processed into the shape of a thin disk with a thickness of 1 mm and a diameter of 14 mm. A fabrication method for a target crystal is described in detail in ref. [5].

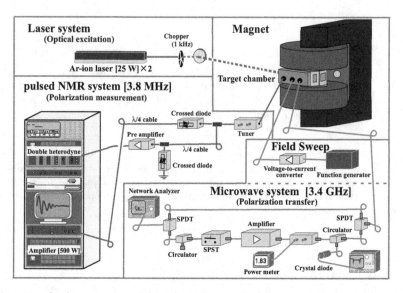

Figure 1. A schematic view of the polarized proton target system. The system consists of a target chamber, a C-type magnet, two Ar-ion lasers, a microwave system and an NMR system.

Figure 1 shows a schematic of the polarized proton target system. The system consists of a target chamber, a C-type magnet that produces an external magnetic field, two Ar-ion lasers for optical excitation, a microwave system and a field sweep system for polarization transfer, and an NMR system for measuring relative value of proton polarization.

A target crystal is placed in the target chamber which is mounted in the center of the C-type magnet.[3] The magnet produces the maximum field of 700 mT and a typical operating field during an experiment is 90 mT. The field uniformity is better than 0.2 mT over the target volume. The uniformity is sufficiently smaller than the internal field of 4 mT in the target crystal.

Two Ar-ion lasers with the total maximum power of 25 W are used to excite pentacene molecules. The laser beams are pulsed by an optical chopper. Typical repetition rate and pulse duration for the optical excitation are 2 kHz and 13 μs, respectively. The resulting average power is 650 mW. The pulsed laser beams are transmitted by optical fibers to the target. Typical transmission efficiency is 50%.

The polarization transfer is carried out with the microwave and field sweep systems. The frequency of microwave is 3.4 GHz which corresponds to an ESR frequency of the lowest triplet state of pentacene in 90 mT. A network analyzer is used as a microwave source. The microwaves from the network analyzer are pulsed by a PIN-diode switch and amplified by a solid-state amplifier, whose maximum output power is 10 W. The amplified microwaves reach the target through a directional coupler and a circulator. The transmitted power to the target is monitored by a power meter, which is attached to the directional coupler with a directivity of 30 dB. The reflected microwaves from the target are detected by a crystal diode which is connected to the circulator. The detected signal is used to tune the microwave frequency.

The relative value of proton polarization during the scattering experiments is measured with a pulsed NMR spectrometer. The NMR frequency in 90 mT is 3.8 MHz. RF pulses from the NMR spectrometer pass through a crossed diode and a tuner, and then reach a 35-turn NMR coil. An NMR signal induced in the NMR coil by RF pulses is passed through the tuner and is amplified with a preamplifier that has a 30 dB gain. The amplified NMR signal is processed with a receiver of double-heterodyne type. Absolute value of proton polarization is calibrated by measuring the analyzing power for $\vec{p}+^4$He elastic scattering at 80 MeV/u.

3. Target performance

The proton polarization during the experiment performed in 2005 is shown in Fig. 2. The polarization build-up was started at 0 days without ^{6}He beam irradiation. The fluctuation of the build-up curve was caused by adjustment for the polarization condition. The proton polarization reached the maximum just before started ^{6}He beam irradiation at 1.8 days. The maximum proton polarization in the experiment was 20.4(58)%. The average proton polarization during the beam irradiation was 13.8(39)%.

The direction of the proton polarization was reversed three times to reduce the systematic uncertainties in the scattering experiment. In the scattering experiment carried out in 2003, the polarization direction was

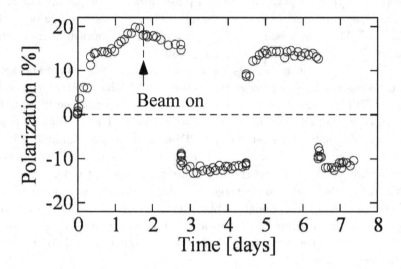

Figure 2. Proton polarization during the scattering experiment. The maximum polarization was 20.4(58)%. After the beam irradiation was started at 1.8 days, proton polarization decreased due to radiation damage. The polarization direction was reversed three times by using a pulsed NMR method.

reversed by build-up to the opposite direction after polarization destruction. This method took approximately 10 hours to recover the polarization. To avoid the waste of time, we have introduced the polarization reversal by using the pulsed NMR method. In this method, the polarization vector rotates around the oscillating field applied perpendicular to the external field. The rotation angle depends on the length of time the oscillating field is on, t, and the strength of the oscillating field, H_1, i.e.

$$\theta = 2\pi\gamma t H_1, \qquad (1)$$

where γ is the gyromagnetic ratio. Figure 3 shows a result of the polarization reversal by the pulsed NMR method in the scattering experiment. We achieved the polarization reversal in $t=2.2$ μs with the reversal efficiency of approximately 70%. By using this method, a scattering experiment can go on without a long interruption for the polarization build-up.

The proton polarization decreases gradually due to radiation damage after the beam irradiation was started, as can be seen from Fig. 2. To estimate the contribution from the radiation damage to the relaxation rate, we measured the relaxation rate before and after the scattering experiment. The relaxation rate measured before the experiment, Γ_{before},

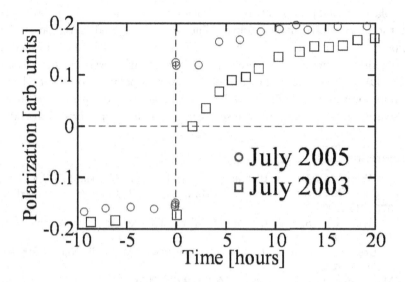

Figure 3. A result of the polarization reversal by using a pulsed NMR method. The polarization direction was reversed in 2.2 μs with the reversal efficiency of 70%. A result of the polarization reversal by a build-up method is also plotted for comparison.

was 0.127(6) h^{-1} and that measured after the experiment, Γ_{after}, was 0.295(4) h^{-1}. The total dose and the spot size of ^{6}He beam were 9.9×10^{10} and 10 mmϕ in FWHM, respectively. To deduce the contribution from the radiation damage, Γ_B, we now assume that Γ_{after} can be written as

$$\Gamma_{\text{after}} = \Gamma_{\text{before}} + \Gamma_L + \Gamma_B . \qquad (2)$$

Γ_L is the relaxation rate caused by laser irradiation and is linearly increased with the laser power and with time for laser irradiation. The proportionality constant obtained from an independent study is (1.1±0.5) ×10^{-3} W^{-1}h^{-2}. For the scattering experiment, Γ_L was estimated to be 0.036(16) h^{-1} using the proportionality constant. ¿From the relaxation rates, Γ_B was obtained as 0.132(17) h^{-1}. The relaxation rate due to radiation damage accounts for 45% of all relaxation rate for the scattering experiment. For a higher beam intensity or a longer experimental run, the contribution from Γ_B will increase. Since the proton polarization decreases with increasing relaxation rate, Γ_B should be reduced periodically by changing the target crystal or by annealing.

4. Summary

A polarized solid proton target for RI beam experiments has been developed at the CNS, University of Tokyo. The target was used in the scattering experiments with ^{6}He beams at RIKEN. In the experiment performed in 2005, the maximum and average proton polarization were 20.4(58)% and 13.8(39)%, respectively. The decrease in proton polarization during beam irradiation was mainly due to radiation damage, whose contribution to the relaxation rate was 45% for the scattering experiment performed in 2005.

References

1. T. Wakui, M. Hatano, H. Sakai, A. Tamii and T. Uesaka, 15th Int. Spin. Physics Symposium, Upton, New York, 2002, AIP Conference Proceedings **675**, 911 (2003).
2. A. Henstra, T.-S. Lin, J. Schmidt, W.Th. Wenckebach, Chem. Phys. Lett. **165**, 6 (1990).
3. T. Uesaka, M. Hatano, T. Wakui, H. Sakai and A. Tamii, Nucl. Instr. and Meth. **A 526**, 186 (2004).
4. M. Hatano, et al., Eur. Phys. J. **A 25**, 255 (2005).
5. T. Wakui, M. Hatano, H. Sakai, T. Uesaka and A. Tamii, Nucl. Instr. and Meth. **A 550**, 521 (2005).

LASER-MICROWAVE DOUBLE RESONANCE SPECTROSCOPY IN SUPERFLUID HELIUM FOR THE MEASUREMENT OF NUCLEAR MOMENTS*

T. FURUKAWA AND T. SHIMODA

Dept. Phys., Osaka Univ., Toyonaka, Osaka 560-0043, Japan
E-mail: take@valk.phys.sci.osaka-u.ac.jp

Y. MATSUO, Y. FUKUYAMA AND T. KOBAYASHI

RIKEN, Wako, Saitama 351-0198, Japan

A. HATAKEYAMA

Dept. of Basic Sci., Univ. of Tokyo, Komaba, Tokyo 153-8902, Japan

T. ITOU AND Y. OTA

Dept. of Phys., Meiji Univ., Kawasaki, Kanagawa 214-8571, Japan

1. Introduction

Nuclear electro-magnetic moments (nuclear moments) are one of the most important quantities in the nuclear structure study because these values are directly correlated with nuclear structures, such as the states and motions of nucleons in the nuclei and the deformation of charge distribution. Therefore, the investigation of nuclear structures by measuring the nuclear moments, particularly those of radioactive nuclei near the drip-line, is an important and fascinating topic in the nuclear physics.

However, measuring the nuclear moments of radioactive nuclei is difficult because of the low yield and high contamination of those nuclei. In addition, there are no useful methods to generate and conserve their spin

---

*This work was supported in part by the Grant-in-Aid for Scientific Research (No. 15654035) by the Ministry of Education, Culture, Sports, Science and Technology, Japan. One of the authors (T. F.) acknowledges the support from the JSPS Research Fellowships for Young Scientists.

polarization of many kinds of short-lived radioactive nuclei, which is essential for nuclear moment measurements. Therefore it is necessary to develop a new, versatile measurement method for the short-lived radioactive nuclei.

We are now developing a new measurement method to determine the nuclear moments by measuring the hyperfine splitting energy of radioactive atoms immersed in superfluid helium (He II) with "laser-microwave double resonance spectroscopy in He II". Our plan is motivated by several fascinating properties of He II. The condensed helium matrices like He II, which are transmissive for light from infrared to ultraviolet, are one of the most noticeable host matrices for optical spectroscopic study in this decade[1]. The atoms implanted in He II have a broadened absorption line spectrum (≥ 10 nm) because of perturbing interactions from the surrounding He atoms. Despite these strong interactions, the spin polarization of atoms immersed in He II is conserved for a long period, as recently confirmed experimentally for Cs atoms (electronic spin relaxation time $T_1 \geq 2$ s) in our study[2]. The broadened line spectrum itself is very promising for optically polarizing a wide variety of atomic species not only alkalis but also non-alkali atoms such as Al and Mg, because the broadened spectrum covers all of complicatedly split lines in many cases. Furthermore, high energy ($\simeq 50$ MeV/u) radioactive nuclear beams produced by nuclear reactions like projectile fragmentations are easily stopped and trapped in the measurement region of He II during the lifetimes of the nuclei[3]. Those properties suggest that He II is a suitable medium for measuring the hyperfine structure of the impurity atoms to determine the nuclear moments with the laser spectroscopic method.

We now report the measurement of hyperfine transition in the ground state of Cs atoms ($|F\rangle : |4\rangle \rightarrow |3\rangle$) immersed in He II with the double resonance method to compare the performance of our system with that of a previously reported result[4].

2. Experiment

2.1. *Double resonance method*

The laser-microwave double resonance spectroscopy is based on the fact that laser-induced fluorescence (LIF) intensity I observed with irradiation of circularly polarized "pumping" light, tuned to the wavelength of the $D1$ excitation line of Cs atoms in He II, is related to the electronic spin polarization of Cs atoms as

$$I \propto N_{\text{Cs}} \times (1 - P_l \times P_z), \tag{1}$$

where N_{Cs} is the number of Cs atoms, P_l is the polarization of pumping laser, and P_z is the electronic spin polarization of the atoms[5,6]. The observed LIF intensity becomes smaller with increasing the spin polarization of the atoms by optical pumping (Fig.1-a and b). The important point is that, if the microwave resonance occurs, the LIF intensity is increased because the electronic spin polarization is destroyed by the hyperfine transition (Fig.1-c).

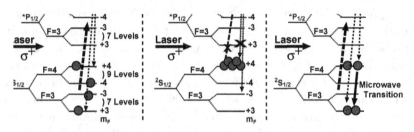

Figure 1. Schematics of the laser-microwave double resonance method:(a) optical pumping; (b) quenching of LIF by complete polarization; (c) re-emission of LIF by depolarization at microwave resonance.

2.2. Details of the measurement

At the beginning of the measurement, Cs atoms are laser-sputtered and introduced with two sputtering lasers[2] (ablation laser for ablating a CsI sample placed above the He II surface and dissociation laser for dissociating CsI clusters introduced in He II) into the optical detection region in He II, where pumping laser light and microwave for the hyperfine transition are both irradiated continuously. The introduced Cs atoms are optically pumped and polarized with irradiation of circularly polarized pumping laser. The LIF intensity is then decreased during a period shorter than 1 ms [2]. Before and after the optical pumping, linearly polarized pumping laser is irradiated to count the number of Cs atoms because the LIF intensity with linearly polarized laser light is related only to the number of Cs atoms [see Eq.(1)]. By sweeping the microwave frequency, we can observe the hyperfine transition from the increase of LIF intensity to determine the hyperfine splitting energy. It is to be noted that each of the hyperfine levels splits into Zeeman sub-levels due to an external magnetic field. Because of this effect, the transition frequencies of $|F, m_F\rangle : |4, +4\rangle \to |3, +3\rangle$ and $|4, -4\rangle \to |3, -3\rangle$, measured with σ^+ and σ^- pumping lasers, respectively, are different. The non-perturbed hyperfine splitting energy is obtained as the average of two resonant frequencies[7].

3. Experimental result

Figure 2 shows the LIF intensity as a function of microwave frequency. With Lorentzian fittings, resonant frequencies are determined preliminarily to be 9.25773(1) GHz in σ^+ pumping and 9.24352(1) GHz in σ^- pumping. Thus the hyperfine coupling constant A of ^{133}Cs atoms immersed in He II is preliminarily determined as 2.312655(5) GHz, which is consistent with the previous data[4] whereas our uncertainty is 20 times smaller than the previous one. This value is slightly larger ($\simeq$0.6 %) than that in vacuum, probably due to the effects of He II pressure. The obtained narrow resonance linewidth of approximately 150 kHz and the resonant frequency close to that in vacuum are suitable for the measurement of the nuclear moments within an uncertainty of 1%, which satisfies the requirement of most nuclear structure investigations. Following this successful measurement of the Cs hyperfine transition in He II, we will investigate the hyperfine structure of ^{85,87}Rb isotopes to confirm that the helium pressure effects work in the same manner for the isotopes with the same electronic structure.

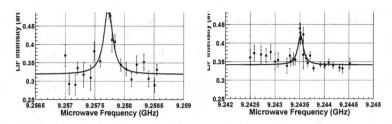

Figure 2. Measured hyperfine transitions with σ^+ (left) and σ^- (right) pumping.

References

1. B. Tabbert et al., J. Low. Temp. Phys. **109**, 653 (1997).
2. T. Furukawa et al., accepted for publication in Phys. Rev. Lett..
3. T. Shimoda et al., Nucl. Phys. **A588**, 235c (1995).
4. Y. Takahashi et al., Z. Phys. **98**, 391 (1995).
5. M. Arndt et al., Phys. Rev. Lett. **74**, 1359 (1995).
6. S. Lang et al., Phys. Rev. **A60**, 3867 (1999).
7. S. Lang et al., Europhys. Lett. **30**(4), 233 (1995).

POLARIZED ^{11}LI BEAM AT TRIUMF AND ITS APPLICATION FOR SPECTROSCOPIC STUDY OF THE DAUGHTER NUCLEUS ^{11}BE

T. SHIMODA, Y. HIRAYAMA, H. IZUMI, Y. AKASAKA, K. KAWAI,
I. WAKABAYASHI, M. YAGI AND Y. YANO
Dept. of Phys., Osaka Univ., Toyonaka, Osaka 560-0043, Japan
E-mail: shimoda@phys.sci.osaka-u.ac.jp

A. HATAKEYAMA
Dept. of Basic Sci., Univ. of Tokyo, Komaba, Tokyo 153-8902, Japan

C.D.P. LEVY AND K.P. JACKSON
TRIUMF, 4004 Wesbrook Mall, Vancouver, BC, Canada V6T 2A3

H. MIYATAKE
Institute of Particle and Nuclear Studies, KEK, Ibaraki 305-0801, Japan

High polarization (50-80%) of nuclear spin has been successfully achieved in low energy alkali radioactive nuclear beams (10-60 keV) by the collinear optical pumping technique at TRIUMF. The polarized ^{11}Li beam was used to study the excited states in ^{11}Be through the β-delayed neutron- and γ-decays. The β-decay asymmetry was very effective to establish the level scheme and decay scheme: The spins and parities of 7 levels in ^{11}Be were firmly assigned for the first time.

1. Introduction

The parity violation in weak interaction, first demonstrated by Wu et al.[1], has opened up fruitful applications of β-decay asymmetry measurements for spin-polarized nuclei to various fields of science. Since the accuracy of asymmetry is inversely proportional to $\sqrt{YP}$, where Y and P are the β-ray yield and the polarization of the parent nucleus, respectively, it is more important to achieve high polarization than high yield. At TRIUMF ISAC, the ISOL-based low-energy (10-60 keV) radioactive nuclear beam facility, highly polarized (50-80%) alkali beams (Li and Na) have been obtained by the collinear optical pumping technique[2]. We have performed spectroscopic measurements of the β-delayed neutron- and γ-decay of polarized ^{11}Li (I^{π}

= $3/2^-$, $T_{1/2}$ = 8.5 ms) to explore the excited states of the daughter nucleus ^{11}Be. Owing to high polarization of ^{11}Li (∼55%), firm spin-parity assignments for seven levels in ^{11}Be have been made for the first time[3].

In the present work, the polarizer is briefly described, and as an example of the successful applications of the polarized beams, the β-delayed decay measurements for polarized ^{11}Li is discussed.

2. Polarizer at TRIUMF ISAC

The TRIUMF polarizer, commissioned in 2001, has been designed with emphasis on achieving both high polarization and high transmission efficiency and on obtaining reionized beams[2]. Figure 1 shows the schematic layout of the polarizer beam line. The radioactive alkali ion beam from the ISOL is neutralized by charge exchange in a Na vapor jet with up to 90% efficiency. The beam is then longitudinally polarized within a distance of 1.9 m (∼2 μsec transit time) by optical pumping on the D_1 transition with counter-propagating circularly polarized light. Next, the polarized beam is reionized in a cold He gas target with 66% efficiency and directed to experiments by the electrostatic optics. The electrostatic bends after the polarizer isolate the experiments from the Na vapour and laser light in the polarizer. If the bends are 90° in total, the polarization with respect to the beam direction is changed from longitudinal to transverse. The overall transmission efficiency is about 60%.

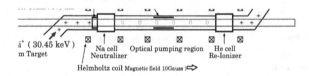

Figure 1. Schematic of the polarizer beam line.

In order to achieve high polarization, two ground-state hyperfine levels, which are separated by 902 MHz for the case of ^{11}Li, must be pumped. For this purpose the laser beam is split in frequency with an EOM (Electro-Optic Modulator) tuned to the required frequency. Also essential for high polarization is to match the laser bandwidth (∼1 MHz) to the Doppler broadening (∼ 150 MHz) of the atomic beam. The broadening is caused by multiple collisions with Na atoms in the neutralizer. The use of another two resonant EOMs in cascade, following the first EOM for frequency splitting, is very effective to produce laser sidebands and broaden the effective laser bandwidth[4].

To date, high polarization is successfully achieved for various Li[2,3] and Na[5] isotopes: ^{8}Li: 80%, ^{9}Li: 56%, ^{11}Li: 55%, ^{20}Na: 57%, ^{21}Na: 56%, ^{26}Na: 55%, ^{27}Na: 51%, ^{28}Na: 45%. The reason for less polarization than that expected from simulation based on the rate-equations is to be clarified.

3. β-delayed decay spectroscopy for ^{11}Li

In order to investigate the level scheme of ^{11}Be, in which the spin-parity assignments are made only for a few low-lying states, the spectroscopic measurements[3] were performed with highly polarized ^{11}Li for the sequential decays $\overrightarrow{^{11}\text{Li}_{\text{g.s.}}} \xrightarrow{\beta} {}^{11}\text{Be}^* \xrightarrow{n} {}^{10}\text{Be}^* \xrightarrow{\gamma} {}^{10}\text{Be}_{\text{g.s.}}$.

The transversely polarized ^{11}Li beam (30.5 keV) was stopped in a Pt foil, which was surrounded by an assembly of β-, neutron- and γ-detectors placed in the atmosphere. For the β-n, β-γ, and β-n-γ coincidences, the β-decay asymmetry was measured by the two β-ray detectors placed in the same and opposite sides with respect to the polarization direction. As in the usual β-delayed decay spectroscopy, the assignments of the level energies and decay paths were made from the discrete energies of delayed neutron and/or γ-rays. The highlight of the present experiment is the unambiguous spin-parity assignments of the daughter states in ^{11}Be through the β-decay asymmetry. The method is based on the fact that the allowed β-decay from a polarized nucleus (polarization P) shows an angular distribution of $W(\theta) = 1 + AP\cos\theta$ ($\theta = 0$ for the polarization direction), which is strongly dependent on the initial and final state spin values through the asymmetry parameter A. In the case of allowed decay of ^{11}Li$_{\text{g.s.}}$ (3/2$^-$), the asymmetry parameter A takes very discrete values of -1.0, -0.4 or $+0.6$, depending on the possible spin-parity of 1/2$^-$, 3/2$^-$ or 5/2$^-$, respectively, of the daughter state in ^{11}Be. The large difference in A is very effective for firm spin-parity assignments.

Figure 2 shows the neutron spectrum (upper panel) and the β-decay asymmetry parameter A in coincidence with the neutrons (lower panel)[3]. The expected values for A are shown by the horizontal lines in the lower panel. It is clearly seen that the asymmetry parameter drastically changes at the neutron peak position and is consistent with the expected value, indicating that the prominent peaks are mostly of a single neutron peak. It is further seen that many neutron peaks are overlapping with each other in the continuum region, and because of cancellation in asymmetry, the observed asymmetry parameters are far from the expected values. Some of the neutron peaks were resolved by examining the β-n-γ coincidence. However, many peaks were left unresolved. The detailed fitting so as to reproduce both the neutron spectrum and the asymmetry parameter resolved the peaks very effectively. Thus, 18 peaks were assigned as shown in Fig. 2 and the spins and parities of the ^{11}Be states emitting these neutrons were determined, and the level scheme and the decay scheme in ^{11}Be have been established[3]. The first firm spin-parity assignments were made for levels; $3/2^-$ ($E_x = 2.69$ MeV); $5/2^-$ (3.89); $3/2^-$ (3.96); $5/2^-$ (5.24); $3/2^-$ (8.02); $3/2^-$ (8.82); $3/2^-$ (10.6).

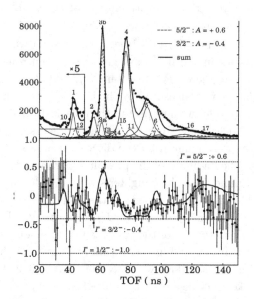

Figure 2. Neutron counts (upper panel) and β-ray asymmetry in coincidence with neutron (lower panel) as a function of time-of-flight. The thin lines (upper panel) are the resolved neutron peaks and the thick line (lower panel) is the average of the asymmetry weighted by the resolved peak intensity.

References

1. C.S. Wu et al., Phys. Rev. **105**, 1413 (1957).
2. C.D.P. Levy et al., Proc. of PST2001 (World Scientific) p.334; Nucl. Phys. **A701**, 253c (2002); Nucl. Instr. and Meth. **B204**, 689 (2003).
3. Y. Hirayama et al., Phys. Lett. **B611**, 239 (2005).
4. A. Hatakeyama et al., Proc. of PST2001 (World Scientific) p.339.
5. K. Minamisono et al., Hyperfine Int. DOI 10.1007/s10751-005-9107-2 (2005).

POLARISED SOLID TARGETS AT PSI: RECENT DEVELOPMENTS

B. VAN DEN BRANDT[1], P. HAUTLE[1], J. A. KONTER[1], F. M. PIEGSA[1,2],
J. P. URREGO-BLANCO[1,3]

[1] *Paul Scherrer Institute, CH-5232 Villigen PSI, Switzerland*
[2] *Physics Department, Technische Universität München,
D-85748 Garching, Germany*
[3] *Dept. Physics and Astronomy, Univ. of Tennessee, Knoxville, TN 37996, USA*

An overview of recent developments and projects at PSI in the field of polarised solid targets is given. They include the realisation of a frozen spin target with a special dilution refrigerator for an experiment on a cold polarised neutron beam, which aims at a precise determination of the doublet nd-scattering length.
Furthermore, we are investigating the possibility of introducing polarisation observables in nuclear spectroscopic studies with radioactive ion beams, at energies around the Coulomb barrier, through thin polystyrene targets.
In the field of target materials, a systematic study of the DNP characteristics of biradicals, dissolved in a matrix of protonated polystyrene, in comparison to the widely used monoradical TEMPO has been performed.

A frozen spin polarised target for experiments on a cold neutron beam

On the cold polarised neutron beam line FUNSPIN at SINQ at PSI an experiment is currently running which aims at a precise determination of the spin dependent doublet nd-scattering length, a low energy parameter particularly well suited to fix three-body forces in novel effective field theories. The incoherent neutron scattering length $a_{i,d}$ can be determined directly using the phenomenon of pseudomagnetic precession. The spin of neutrons passing through a polarised target precesses around the axis of nuclear polarisation with the precession angle being proportional to the incoherent scattering length a_i of the nuclear species present in the sample. [1] The angle can be measured very accurately using Ramsey's well-known atomic beam technique, adapted to neutrons. [2]

In order to circumvent severe limitations in accuracy, imposed by having to know absolute values of polarisation, density or thickness of the target,

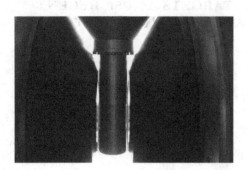 

Figure 1. Left: Outer vacuum tube of the dilution refrigerator fitting tightly between the pole pieces of the electro-magnet. Right: Teflon holder positioning the 13×7×1.5mm slabs of deuterated polystyrene inside the ^{4}He filled target cell.

we adopt a measurement scheme where the incoherent scattering length of the deuteron is determined relative to the one of the proton, [1] which is known to high accuracy. This requires that the precession angles from the two hydrogen isotopes present in the target sample have to be measured separately, i.e. the polarisation of one species is destroyed whereas the other is kept frozen and vice versa. As a consequence the sample temperature has to be lower than 100 mK to keep cross relaxation processes negligible during the precession measurements. We thus realised a frozen spin type polarised target system which furthermore is specially adapted for the use on a cold neutron beam. The ^{3}He-^{4}He dilution refrigerator is based on the PSI design [3] with a cooling power of 1 mW at 100 mK and a base temperature of 50 mK. In order to avoid that the beam has to pass through the strongly absorbing ^{3}He, the target is contained in a separate cell filled with ^{4}He and is cooled by the mixing chamber via a sintered silver heat exchanger. The left part of Figure 1 shows the cryostat, just fitting between the pole pieces of the magnet. The holder with an appropriate collimator positioning the sample in the ^{4}He cell is seen on the right. The cryostat has been in operation on beam during a period of 3 month in summer 2005. Two different samples of ∼95% deuterated polystyrene doped with deuterated TEMPO have been polarised to values optimised for the Ramsey experiments ($P_D \sim 20\%$). The polarisation has been frozen at temperatures of ∼80 mK.

Thin polarised targets for RI beam experiments

At low energies, stable polarised beams are used for spectroscopic purposes. It was proposed to extend these types of experiments to nuclei far from stability by using radioactive ion beams (RIBs) and polarised targets.[4] For this purpose we are developing a solid polarised proton target in the thickness range between 20 μm and 100 μm based on a polymeric foil. Such a target would be a useful tool in the determination of excitation functions in resonant reactions, in studies of one nucleon transfer reactions as well as in probing the matter density of atomic nuclei.[4,5]

Several problems have to be addressed for the realisation of such a polarised target system:

(a) the effect of the magnetic field on the trajectory of the recoiling particles as well as on the performance of the detection system positioned in the field close to the target

(b) the design and coupling of a cryogenic system to a vacuum line, i.e. the amount of material on the beam path should be minimised

(c) the influence of beam heating and radiation damage on the target polarisation which can only be thoroughly studied under real operating conditions.

A cryogenic system similar to the one described in the previous section is under construction. A dilution refrigerator provides a base temperature low enough to freeze the polarisation at a magnetic field sufficiently low (B < 1 T) to minimise effects on the particle trajectories. The sample foil is mounted in a separate cell on a copper frame that is in thermal contact with the mixing chamber of the dilution refrigerator. A superfluid ^{4}He film assures a good overall cooling. The target cell, directly in the beam vacuum, should have maximum transparent beam windows. A candidate material is Si_3Ni_4, which has an extremely low energy loss: < 10 keV for 10 MeV protons and < 270 keV for 60 MeV ^{12}C ions at a thickness of 300 nm.

Tests of the different components of the target system are in progress. A proof of principle test of the system is planned to be performed at the Philips cyclotron at PSI in the near future. We will use the reaction $\vec{p}-^{12}C$ with a carbon beam in inverse kinematics.

Dynamic nuclear polarisation with biradicals

The measurement of distances on the nanometer scale (1.5 - 8 nm) is still a challenging task for EPR (electron paramagnetic resonance) methods. Recently shape-persistent biradicals have been synthesised by Godt et al. [6] to explore the potential of new pulse sequences. The two nitroxide radicals at a fixed distance at the end of a rod like molecule offer the unique possibility to study with SANS and dynamic nuclear polarisation (DNP) [7] the evolution of the nuclear spin polarisation of the protons along the axis between the two unpaired electrons. [8] The observation that a biradical, dissolved in a 98% deuterated matrix of polystyrene, very efficiently supports the DNP process of protons, has motivated a more systematic study of the DNP characteristics of biradicals in comparison with the widely used monoradical TEMPO. A series of samples (14×14×2 mm) with decreasing free electron concentration has been prepared by dissolving protonated polystyrene in toluene, containing the radical, followed by drying and hot pressing the obtained thin films. The DNP experiments were performed in a continuous flow ^{4}He refrigerator at 1.2 K and 3.5 T.

Figure 2 shows preliminary results for the proton polarisation obtained with a biradical with an intraradical distance of 3.68 nm for three concentrations of unpaired electrons in the sample (4.3×10^{19}e$^-$/cm^3, 1.44×10^{19}e$^-$/cm^3, 4.8×10^{18}e$^-$/cm^3). The results of the samples doped with TEMPO to the equivalent electron spin concentration are shown in open circles. Except for the largest electron concentration, a significantly higher polarisation can be achieved with the biradical: for the low concentration the biradical yields an 11 times, and for the medium concentration a 2.5 times higher polarisation than TEMPO. Similar observations have been made at T = 90 K. [9] The mean distance between two unpaired electrons in the 4.3×10^{19}e$^-$/cm^3 doped TEMPO sample is comparable to the intraradical distance in the biradical doped sample, thus giving similar e$^-$-e$^-$ dipole couplings. As expected, pure thermal mixing [7] is observed as DNP mechanism and comparable polarisations can be achieved with both types of dopants. When the total number of unpaired electrons is reduced, the e$^-$-e$^-$ cross relaxation might not be efficient enough to assure thermal equilibrium within the electron non-Zeeman reservoir (EnZ), all the more than an inhomogeneous broadening of the EPR line is expected at the present magnetic field. Microwave irradiation then leads to hole burning in the EPR line and not to dynamic cooling of the EnZ. The maximum achievable polarisation should drop and is observed to be close to zero in the sample with the lowest TEMPO concentration. In the case of a biradical doped sample, even though the interradical distances increase, there are still pairs of interacting electrons, that can relax a nuclear spin in a three-spin process, thus supporting the DNP process.

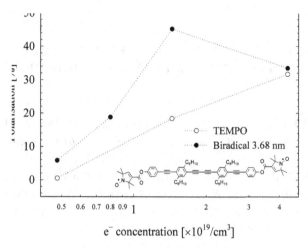

Figure 2. Maximum proton polarisations P achieved with TEMPO and a biradical as a function of the unpaired electron concentration at a temperature of 1.2 K in a field of 3.5 T. Assuming a thermal equilibrium polarisation of 0.3%, the corresponding enhancement ε is given by $\varepsilon = P/0.3$ and is more than 150 in the best case.

Further experiments are in progress to clarify the tentative explanation for the DNP process with biradicals.

Acknowledgement: The biradical has been kindly provided by A. Godt, Universität Bielefeld, Germany.

References

1. B. van den Brandt, H. Glättli, H.W. Grießhammer, P. Hautle, J. Kohlbrecher, J.A. Konter, O. Zimmer, *Nucl. Instr. Meth. A* **526**, 91 (2004).
2. A. Abragam, M. Goldman, *Nuclear magnetism: order and disorder* (Clarendon Press, Oxford, 1982).
3. B. van den Brandt, J. A. Konter, S. Mango, *Nucl. Instr. Meth. A* **289**, 526 (1990).
4. J. P. Urrego-Blanco, B. van den Brandt, E. I. Bunyatova, A. Galindo-Uribarri, P. Hautle, J. A. Konter, *Nucl. Instrum. Meth. B* **241**, 1001 (2005).
5. K. Amos, S. Karataglidis, J. Dobaczewski, *Phys. Rev. C* **70** 024607 (2004).
6. A. Godt, C. Franzen, S. Veit, V. Enkelmann, M. Pannier, G. Jeschke, *J. Org. Chem.* **65**, 7575 (2000).
7. A. Abragam and M. Goldman, *Rep. Prog. Phys.* **41**, 395 (1978).
8. B. van den Brandt, H. Glättli, P. Hautle, J. Kohlbrecher, J. A. Konter, A. Michels, H. B. Stuhrmann, O. Zimmer, PSI Scientific Report 2004.
9. K-N. Hu, H-H. Yu, T. M. Swager, R.G. Griffin, *J. Am. Chem. Soc.* **126**, 10844 (2004).

PRODUCTION OF SPIN-ORIENTED UNSTABLE NUCLEI VIA THE PROJECTILE-FRAGMENTATION REACTION

H. UENO, D. KAMEDA, A. YOSHIMI, AND T. HASEYAMA
RIKEN, 2-1 Hirosawa, Wako, Saitama 351-0198, Japan

K. ASAHI,* M. TAKEMURA, G. KIJIMA, K. SHIMADA, D. NAGAE,
M. UCHIDA, T. ARAI, S. SUDA, K. TAKASE, AND T. INOUE
Department of Physics, Tokyo Institute of Technology, Meguro-ku, Tokyo 152-8551, Japan

T. KAWAMURA
Department of Physics, Rikkyo University, Nishi-Ikebukuro 3-34-1, Toshima-ku, Tokyo 171-8501, Japan

Based on the β-NMR method with the spin-polarized radioactive-isotope beams produced in the projectile-fragmentation reaction, we have recently carried out experiments at RIKEN to measure nuclear moments of neutron-rich aluminum isotopes. In the measurements, the magnetic dipole moments of ^{30,32}Al and the electric quadrupole moments of ^{31,32}Al have been determined. No deviation was found for the obtained magnetic moments of aluminum isotopes in the comparison with conventional shell-model calculations. Also, the obtained quadrupole moments are found to be small, indicating their spherical shapes. These results seem to suggest that these aluminum isotopes are located outside the "island of inversion".

1. Introduction

It has been revealed that spin-oriented radioactive-isotope beams (RIBs) can be produced in the projectile-fragmentation (PF) reaction [1,2], which offers us the opportunity of studying on the structures of nuclei far from the stability line, through the measurement of electromagnetic nuclear moments. Up to now, 14 new magnetic (μ) moments and 6 new quadrupole (Q) moments have been determined in the light unstable nuclear region

---

*Also at RIKEN, 2-1 Hirosawa, Wako, Saitama 351-0198, Japan

at RIKEN by two groups. In our research, neutron-rich p-shell nuclei have been studied so far. The obtained experimental nuclear moments have been shown quite effective in discussing the effect of neutron excess on their nuclear structure, where we discussed the deviation of μ moments from the Schmidt value [3] and the isospin dependence of the effective charges [4]. In order to extend the observation into the neutron-rich sd-shell region, we have recently measured ground-state nuclear moments of neutron-rich aluminum isotopes. In this paper, we first introduce several methods to produce spin-oriented RIBs, then report on our recent result obtained for the aluminum isotopes.

2. Production of spin-oriented radioactive-isotope beams

There are currently two methods for producing RIBs. The isotope separator on-line (ISOL) technique was first applied to RIB production, where a high-energy driver beam of stable light ions is used to bombard a heated target. Then, RI atoms produced in the target through the target fragmentation, fission, and spallation are extracted chemically for post acceleration. The produced RIBs in this method are characterized by their low energy with high quality. Unfortunately, the migration process of RI atoms in the target, however, is so slow that many of the unstable RI may decay before the post acceleration. Furthermore, the available elements of RI are limited in this scheme, since the migration process is governed by chemical properties. Alternatively, an in-flight RIB production method has been developed, where RIBs are produced as the fragment of high-energy heavy-ion beams in the PF reaction. Since RIBs are produced and isotope-separated in a short time, short-lived RIBs can be produced. Thus, the produced RIBs are characterized by their higher energies but greater energy variation and divergence. The two methods have been considered complementary.

In the case of ISOL-based RIBs, the optical-pumping method is suitable for the production of spin polarization because of the small momentum spread. In this method, the atomic polarization produced by injecting circularly-polarized laser light, is then transferred to the nuclear polarization through the hyperfine interaction. Application of this method to PF-based RIBs, however, is not effective because of the large momentum spread. Instead, spin-orientation phenomena found in the PF reaction [1,2] have been utilized. The mechanism of the fragment-induced spin orientation is essentially related to the fact that a portion of the projectile to be removed through the fragmentation process has non-vanishing angular momentum due to the internal motion of nucleons. This mechanism is confirmed in the measurement of spin alignment at high energies [6] and in the systematic measurements of spin polarization [5]. Also, it is interesting to note that recently large spin polarization was found in the single nucleon pick-up reaction at high energies [13].

3. Measurements of nuclear moments

Many ground-state nuclear moments of unstable alkaline isotopes have been measured at ISOLDE CERN based on the optical pumping technique and the laser spectroscopy [7]. The nuclear structure of neutron-halo nuclei was discussed using the obtained experimental nuclear moments of ^{11}Li [8,9] and ^{11}Be [10]. The nuclear moments of ^{31}Na [7] and ^{31}Mg [11] have also been measured. They are important as the nuclei located in the "island of inversion" [12] (This issue is taken up in the next section).

This method, however, is limited to the alkaline isotopes due to the atomic properties. Alternatively, ground-state nuclear moments have been measured utilizing the spin polarization produced in the PF reaction. To determine the nuclear moments, the β-ray detected nuclear magnetic resonance (β-NMR) technique is adopted, where NMR is observed through a change in the β-ray asymmetry: when the polarization P is altered due to the resonant spin reversal, a change appears in the β-ray up/down ratio. Based on this method, the measurement of nuclear moments has been carried out in several facilities.

Spin-oriented PF-based RIBs are also useful to determine the μ moment of isomer states in unstable nuclei, since both the production of spin alignment and the population of isomer states are possible at the same time. The measurement of μ moments can be carried out with the time dependent perturbed angular distribution (TDPAD) method. Thus, the μ-moment measurements of the isomer states in ^{69}Cu, ^{67}Ni [14], and ^{61}Fe [15] have been carried out at GANIL.

It is worth a mention that recently μ moments of short-lifetime excited states ($\tau \sim$ ps) in unstable nuclei have been measured based on the transient field (TF) method [16]. TF is an effective magnetic field acting on ions passing through ferromagnetic foils. Because of large TF strengths nuclear moments can tilt in several degrees before their γ decay, which causes a observable shift in the γ-ray angular distribution. The first measurement using an RIB was carried out at LBL [17]. Development of a detector setup for PF-based RIBs has also been developed at RIKEN [18].

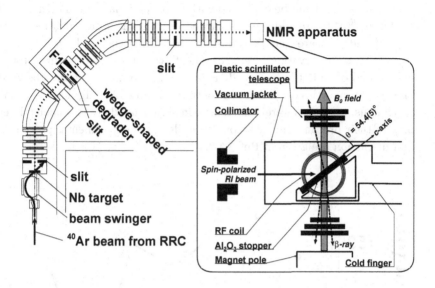

Figure 1. Arrangement of the RIKEN projectile fragment separator RIPS for the production of spin-polarized $^{30-32}$Al beams, and the schematic layout of the β-NMR apparatus.

4. Ground-state nuclear moments of neutron-rich aluminum isotopes at RIKEN

Neutron-rich aluminum isotopes are located near the border of the "island of inversion", where ground-state deformation is observed only in a localized area of the neutron-rich sd-shell region around the neutron number $N = 20$. The presence of the "island of inversion" is considered as manifestations of the deformation induced by the inversion of amplitudes between sd-normal and pf-intruder configurations [12]. Reported several intriguing phenomena are discussed in association with this "inversion" [19,20,21,22,23]. Their nuclear moments allow us to discuss microscopically the evolution of the "inversion".

Experiments were performed using RIKEN projectile-fragment separator RIPS [24]. In the experiments, beams of ^{30,31,32}Al were obtained from the fragmentation of ^{40}Ar projectiles at an energy of $E = 95$ A MeV on a ^{93}Nb target. Thus, μ moments of ^{30,32}Al [25] and Q moments of ^{31,32}Al [26] have been measured successfully. Arrangement of RIPS for the production of spin-polarized $^{30-32}$Al beams and the β-NMR apparatus are schematically shown in Fig. 1. From the obtained experimental data, we found that the μ moment of ^{32}Al can be explained with a conventional shell model as well as that of ^{30}Al [25]. In addition, as shown in Fig. 2, the obtained Q moments of ^{31}Al and ^{32}Al are found to be small, indicating their spherical nuclear shape. These observations seem to indicate that these aluminum isotopes are located outside the border. It is interesting to note that the observed "normal" properties of ^{32}Al are in contrast to those of ^{31}Mg, for which the anomalous properties were reported [11], in spite that these two nuclei differ only in their proton number by one. Details of the experimental procedure and discussion are given in Ref. [25,26].

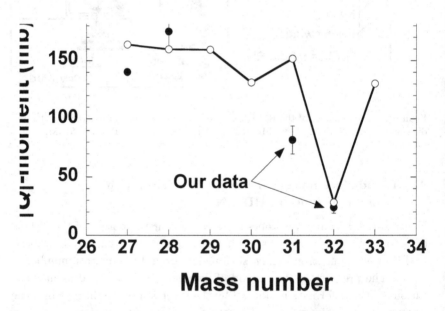

Figure 2. Mass number dependence of nuclear electric quadrupole moments of aluminum isotopes. The open circles show the result of shell model calculations. The solid circles are experimental data, in which present our data of $^{31, 32}$Al are also plotted.

5. Summary

Taking advantage of spin-polarized RIBs from the PF reaction, we have recently carried out experiments at RIKEN to measure nuclear moments of neutron-rich aluminum isotopes. In the measurements, the magnetic dipole moments of ^{30,32}Al and the electric quadrupole moments of ^{31,32}Al have been determined. No deviation was found for the obtained magnetic moments of aluminum isotopes in the comparison with conventional shell-model calculations. Also, the obtained quadrupole moments are found to be small, indicating their spherical shape. These observations suggest that drastic change of the nuclear structure takes place between magnesium and aluminum. Further discussion will be made in the forthcoming paper.

References

1. K. Asahi et al., Phys. Rev. C **43**, 456 (1991)
2. K. Asahi et al., Phys. Lett. B **251**, 488 (1990)
3. H. Ueno et al., Phys. Rev. C **53**, 2142 (1996)
4. H. Ogawa et al., Phys. Rev. C **67** 064308 (2003)
5. H. Okuno et al., Phys. Lett. B **335**, 29 (1994)
6. W.-D. Schmidt-Ott et al., Z. Phys. A **350**, 215 (1994)
7. G. Huber et al., Phys. Rev. C **18**, 2343 (1978)
8. E. Arnold et al., Phys.Lett. B **197**, 311 (1987)
9. E. Arnold et al., Phys. Lett. B **281**, 16 (1992)
10. W. Geithner et al., Phys. Rev. Lett. **83**, 3792 (1999)
11. G. Neyens et al., Phys. Rev. Lett. **94**, 022501 (2005)
12. E.K. Warburton, J.A. Becker, B.A. Brown, Phys. Rev. C **41**, 1147 (1990)
13. D.E. Groh et al., Phys. Rev. Lett. **90**, 202502 (2003)
14. G. Georgiev et al., Eur. Phys. J. A **20**, 93 (2004)
15. I. Matea et al., Phys. Rev. Lett. **93**, 142503 (2004)
16. A.D. Davis et al., Bull. Am. Phys. Soc. **50**, 69 (2005)
17. K.-H. Speidel et al., Eur. Phys. J. A **25**, 203 (2005)
18. H. Ueno et al., Hyp. Int. **136/137**, 211 (2001); A. Yoshimi et al., Nucl. Phys. A **738**, 519 (2004).
19. C. Thibault et al., Phys. Rev. C **12**, 644 (1975)
20. C. Détraz et al., Nucl. Phys. A **394**, 378 (1983)
21. C. Détraz et al., Phys. Rev. C **19**, 164 (1979)
22. D. Guillemaud et al., Nucl. Phys. A **426**, 37 (1984)
23. T. Motobayashi et al., Phys. Lett. B **346**, 9 (1995)
24. T. Kubo et al., Nucl. Instr. Meth. B **70**, 309 (1992)
25. H. Ueno et al., Phys. Lett. B **615**, 186 (2005)
26. D. Kameda et al., *these proceedings*

TILTED FOIL NUCLEAR POLARIZATION

GVIROL GOLDRING
Particle Physics Department,
Weizmann Institute of Science,
Rehovot 76100, Israel
E-mail: gvirol.goldring@weizmann.ac.il

The possibility of polarizing ions in a beam by passing them through foils tilted to the beam axis was first raised by Fano and Macek in a review paper on ionic beam foil spectroscopy [1]. The matter was then taken up by H. G. Berry et al. and they found unexpectedly large circular polarization of the light emitted by the ions issuing from the tilted foil. This discovery led in the following decade to widespread activity in related fields of atomic physics. The newly discovered process aroused great interest in the Weizmann Institute nuclear physics group where hyperfine interactions in randomly oriented ions in vacuum and gas were used extensively to measure nuclear g-factors. However only the absolute value of g could be measured in that way. The tilted foil polarization discovery eventually made it possible to transfer the polarization to the nucleus via vacuum hyperfine interaction and to measure the sign of the nuclear g-factors. A measurement carried out at the Weizmann Institute on ^{16}O(3^-) established mrad angular shifts in the angular distribution of γ's emitted by a nucleus inside an ion issuing from the target tilted to the velocity vector of the ions [2]. The results of these measurements are shown in Fig.1. The 3^- level was excited in the reaction ^{19}F(p,$\alpha\gamma$)^{16}O(3^-) and θ is the angle between the γ's and the p beam.

The subsequent activity in this field was greatly aided by the introduction of multifoil stacks: a number of foils introduced between a target and a stopper with an adjustable distance between them. The assembly was rotatable about the $\mathbf{n} \times \mathbf{v}$ axis in order to obtain the desired tilt angle and to shift the tilt angle from left to right (and the polarization from up to down) Fig.2 shows results of measurements for ^{84}Sr (8_1^+), $T_{1/2}$=163ps and ^{80}Se(2_1^+),$T_{1/2}$=8.5 ps compared with computed and g factor fitted curves[3]. The time through the stack is plotted in units of the mean hy-

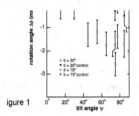 Figure 1

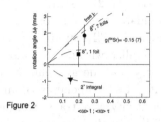 Figure 2

perfine period. ^{80}Se(2_1^+) decays inside the stack even at the shortest target to stopper distance and the only relevant time parameter is the mean life τ. ρ is the double ratio for detectors right and left, and polarization up and down.

Early in the 1980's interest grew in high spin nuclear isomers and in their quadrupole moment Q. In some cases, in particular in the $(49/2)^+$ isomer in ^{147}Gd, it was important to know the sign of Q which was theorized to be negative but there were no tools available for such measurements. We then embarked on a project to adapt the multifoil technique to sign of Q measurements. In this scheme the nuclei are polarized in the foil stack and imbedded in a single crystal with the polarization along the c- axis [4]. The time evolution of the angular distribution of the γ's from the decay of the isomer is then determined algebraically by the parameter qQ, where q is the gradient of the **E** field.

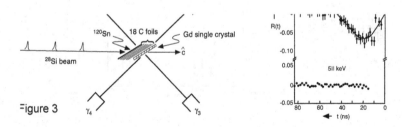 Figure 3

The detector and crystal arrangement is shown in Fig.3 together with R(t): the evolution in time, with respect to the beam pulses of the ratio of two detectors on opposite sides of the beam, for an isomer in ^{144}Gd. The sign of Q of the $\frac{49}{2}^+$ isomer in ^{149}Gd was confirmed to be negative.

With the availability of a working method of nuclear polarization, an attempt was also made to detect and measure a possible parity violation in the $(17/2)^-$ isomer of ^{93}Tc . The only established cases (beyond two standard deviations) of parity violation in bound nuclear states were the 8^+ isomer in ^{180}Hf and the $\frac{1}{2}^-$ state in ^{19}F. The ^{93}Tc isomer was considered a possibility, with polarized isomeric nuclei and an expected 0° – 180° asymmetry of the decay gammas along the polarization direction. Measurements with tilted foil polarized isomers were started in Rehovot and, for improved sensitivity were transferred to GSI, to LNL in Legnaro and them back to GSI[5]. In all these latter measurements a ^{93}Tc isomer beam was prepared, velocity or mass analyzed and passed through a polarizing foil stack to a Pb stopper. The decay γ's were viewed in two Ge detectors aligned with and against the polarization. The level scheme of ^{93}Tc is shown in Fig.4 and the gamma spectrum in Fig.5. There are several isomer lines, other than the 751 keV of the $(17/2)^- \longrightarrow (13/2)^+$ transition in the spectrum, most of them from the ^{93}Tc isomer. They all (leaving out the 751 keV line) exhibit small and consistent anisotropies and so provide an excellent control basis. The anisotropy of the 751 keV line is measured against the average of the other isomer lines. The results of the four measurements conducted are shown in Fig.6 as triple ratios: two detectors, polarization up and down and the 751 keV against the other lines. The first result suggested an anisotropy of about the expected value but the last, with the best statistics, is consistent with zero, and the overall anisotropy is $8(4) \times 10^{-4}$ with an uncertainty too large for the result to be considered significant.

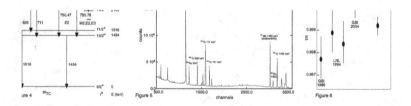

The most recent application of tilted foils was in g-factor measurement in mirror pair partners in the s-d shell. The first measurement was for ^{23}Mg, one of the last T=$\frac{1}{2}$ mirror pairs to be measured. A recent measurement concerned ^{17}Ne, T=$\frac{3}{2}$ I$^\pi = \frac{1}{2}^-$ [2] . Those measurements were carried out at ISOLDE, CERN where copious beams of radionucleides can be produced. The g-factors were measured by NMR in the H-field of a superconducting coil. The polarization serves to establish a 0° - 180° asymmetry of the decay γ's with respect to the polarization. The acceleration voltage at ISOLDE is 60 kV, too low to allow the ions in the beam to pass through foils. The whole apparatus - tilted foil polarizer, cold Pt stopper, He cryostat, detectors - was therefore mounted on a -250 kV platform to accelerate the ion beams to a total of 310 keV. The set up and the measured NMR trace are shown below. The g-factor of ^{17}Ne was established as: g = 1.48(6).

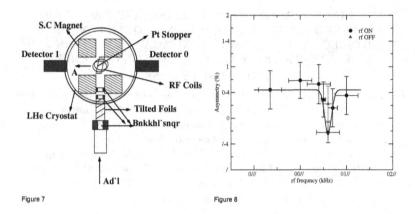

Figure 7   Figure 8

A post accelerator, REX ISOLDE, is now being set up at the ISOLDE laboratory. It will provide, in the first stage, a voltage of 2 MV and higher voltages later. The tilted foil - NMR assembly will be moved to the new locality and even the first stage will constitute an important improvement in the tilted foil context. In the near future this is likely to be the major application of nuclear tilted foil polarization.

References

1. V. Fano et al. Impact Excitation and Polarization of the Emitted Light. Rev. Mod. Phys. **45**,553 (1973).
2. G. Goldring et al. Magnetic hyperfine rotation of a γ-ray angular distribution due to target tilting. Phys. rev. Lett. **38**, 221 (1977).
3. C. Broude et al. The g-factor of the 8^+ Yrast level in ^{84}Sr via a tilted multi-foil experiment. Phys. Lett. **105B**, 119 (1981).
4. E. Dafni et al. Oblate deformation of high spin levels in Gd isotopes. Phys. Rev. Lett. **53**, 2473 (1984).
5. B.S. Nara Singh et al. Parity non conservation in the γ decay of polarized $17/2^-$ isomers in ^{93}Tc. PRC **72** 027303 (2005).
6. L.T. Baby et al. The magnetic moment of the ground state of the T=$\frac{3}{2}$ ^{17}Ne nucleus with tilted foil polarization. J. Phys. **G 30**, 519 (2004).

SPIN POLARIZATION OF ^{23}Ne PRODUCED IN HEAVY ION REACTIONS

M. MIHARA, R. MATSUMIYA, K. MATSUTA, T. NAGATOMO*,
M. FUKUDA, T. MINAMISONO[1], S. MOMOTA[2], Y. NOJIRI[2],
T. OHTSUBO[3], T. IZUMIKAWA[4], A. KITAGAWA[5], M. TORIKOSHI[5],
M. KANAZAWA[5], S. SATO[5], J. R. ALONSO[6], G. F. KREBS[6]
AND T. J. M. SYMONS[6]

Department of Physics, Osaka University, Toyonaka, Osaka 560-0043, Japan
[1]*Fukui University of Technology, Fukui 910-8505, Japan*
[2]*Kochi University of Technology, Tosayamada, Kochi 782-8502, Japan*
[3]*Faculty of Science, Niigata University, Niigata 950-2181, Japan*
[4]*Radioisotope Center, Niigata University, Niigata 951-8510, Japan*
[5]*National Institute of Radiological Sciences, Chiba 263-8555, Japan*
[6]*Lawrence Berkeley Laboratory, CA94720, USA*
E-mail: mihara@vg.phys.sci.osaka-u.ac.jp

Spin polarization of β-emitting nucleus ^{23}Ne ($I^\pi = 5/2^+$, $T_{1/2} = 37.24$ s) produced in heavy ion reactions was observed by means of the β-NMR method. The ^{23}Ne nuclei were produced via projectile fragmentation and single neutron pickup from a Be target using 100A-MeV ^{26}Mg and ^{22}Ne beams, respectively, and then, were implanted into a NaF single crystal at 15 K. Polarization was obtained as $-0.71(17)\%$ and $+2.4(4)\%$ for the fragmentation and pickup reactions, respectively. The positive polarization for the pickup process is consistent with the simple kinematical model.

1. Introduction

Techniques for producing spin polarized radioactive nuclear beams are extremely useful for studying structures of unstable nuclei through measurements of their electromagnetic moments, fundamental interactions observed in nuclear β decay, and material science through the hyperfine interactions. Among them, the projectile fragmentation process in heavy ion reactions at intermediate energy is known as one of the most useful methods and the mechanisms of the polarization production have been studied

---

*Present address: Institute of Physics, University of Tsukuba, 1-1-1, Tennoudai, Tsukuba, Ibaraki 305-8571, Japan.

systematically[1]. Recently, spin polarization of ^{37}K produced through the single proton pickup reaction has been observed using a 150A-MeV ^{36}Ar beam, which suggests that the nucleon pickup process might be also available to produce polarized radioisotope beams[2]. In the present experiment, we have attempted both projectile fragmentation and nucleon pickup processes to generate spin polarization of β-emitting nucleus ^{23}Ne ($I^\pi = 5/2^+$, $T_{1/2} = 37.24$ s). The degree of polarization of ^{23}Ne obtained through these two different processes was measured by means of the β-NMR method.

2. Experimental

The present experiment was performed at Heavy Ion Medical Accelerator in Chiba (HIMAC) in National Institute of Radiological Sciences (NIRS). The experimental method is similar to the previous one[3]. The projectile fragmentation of ^{26}Mg and the single neutron pickup by ^{22}Ne were applied to produce ^{23}Ne using 100A-MeV ^{26}Mg and ^{22}Ne beams with intensities of 1.2×10^9 pps and 2.1×10^9 pps, respectively, impinged on a 2-mm thick Be target. The ^{23}Ne nuclei were separated by using the secondary beam course SB2[4]. The ejection angle of $1.0° \pm 0.6°$ and the momentum were selected to induce polarization of ^{23}Ne. Thus obtained polarized ^{23}Ne nuclei were slowed down by an energy degrader to implant into a cubic single crystal NaF catcher cooled down to 15K placed in a magnetic field of 1 T. The degree of polarization P for ^{23}Ne in NaF was deduced through observation of the asymmetry AP in the β-ray angular distribution $W(\theta) \approx 1 + AP cos(\theta)$, which was detected by two sets of plastic-counter telescopes located above ($0°$) and below ($180°$) the catcher relative to the polarization direction. The asymmetry parameter A averaged over branches for the β decay of ^{23}Ne is -0.75[5]. Beta rays were counted in the beam-off time for 80 s after beam irradiation time of 38 s. Beta-ray yield was typically 300−400 cps for the both reactions. Polarization was inverted through the adiabatic fast passage, by applying a frequency modulated rf magnetic field ($\sim$2.5 mT) with the frequency of 3293 ± 100 kHz for 10 ms between the beam-on and off time. Polarization was deduced from the asymmetry change between the rf-on and off cycles which were repeated alternately.

3. Results and Discussion

The polarization and yield of ^{23}Ne for the two processes are shown in Figure 1 as a function of the relative momentum $\Delta p/p_0$ normalized by the one at the beam velocity. The polarization of ^{23}Ne was successfully maintained

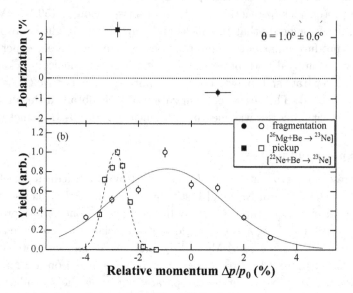

Figure 1. (a) Polarization and (b) yield of ^{23}Ne plotted as a function of the relative momentum normalized by the one at the beam velocity.

for a long time in NaF at 15 K as shown in Figure 2. The spin-lattice relaxation time T_1 was deduced as $T_1 = (140 ^{+100}_{-40})$ s. The width of the momentum distribution of projectile fragments is ~200 MeV/c, which is near the prediction by the Goldhaber model[6]. In contrast, the momentum distribution of pickup products is quite narrow and its center is considerably below the one at the beam velocity as $\Delta p/p_0 = -2.9\%$. In the case that the ground states of ^{23}Ne and ^{8}Be are produced by the direct reaction ^{9}Be(^{22}Ne, ^{23}Ne)^{8}Be, $\Delta p/p_0 = -2.3\%$ is predicted. The slightly lower value of $\Delta p/p_0$ for the experimental distribution might be due to the excitation of the reaction products or the direct breakup of the target residue like 2α. Polarization of ^{23}Ne was measured at the high momentum side for the fragmentation and at the peak for the pickup reaction, which gave values of P as $-(0.71 \pm 0.17)\%$ and $+(2.4 \pm 0.4)\%$, respectively. The signs of polarization for the both reactions are consistent with the previous results in the case of light mass target[1,2], which is explained by the far-side trajectory and the velocity mismatch between the projectile and the removal (picked up) nucleon(s). The fact that much larger polarization was obtained for the pickup reaction than that for the fragmentation reaction implies that the population of the excited states ^{23}Ne* might be suppressed for the pickup process, which should cause depolarization of the ground state of ^{23}Ne especially by way of the $1/2^+$ state of ^{23}Ne*(1st, $E_x = 1017$ keV)[5].

Although further experimental studies are desired to clearly understand the mechanism of the polarization in the pickup reactions, the present results strongly support that the nucleon pickup reactions have much potential to produce polarized unstable nuclei.

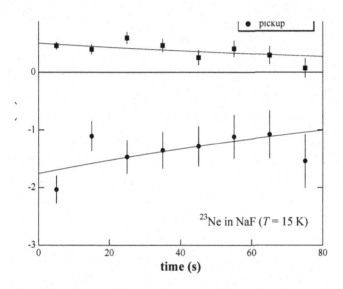

Figure 2. Relaxation of the polarization of ^{23}Ne in NaF at 15 K.

References

1. H. Okuno et al., *Phys. Lett.* **B335**, 29 (1994).
2. D. E. Groth et al., *Phys. Rev. Lett.* **90**, 202502 (2003).
3. K. Matsuta et al., *N. Phys.* **A701**, 383c (2002).
4. M. Kanazawa et al., *N. Phys.* **A746**, 393c (2004).
5. R. B. Firestone et al., *Table of Isotopes*, CD-ROM, 8th Edition (1996).
6. A. S. Goldhaber et al., *Phys. Lett.* **B53**, 306 (1974).

New Methods

SPIN-POLARIZATION USING OPTICAL METHODS

TAKASHI NAKAJIMA

Institute of Advanced Energy, Kyoto University
Gokasho, Uji, Kyoto 611-0011, Japan
E-mail: t-nakajima@iae.kyoto-u.ac.jp

We theoretically consider three different schemes to produce spin-polarized electrons/ions with some discussions and results for experimental realization. Our schemes involve neither optical pumping nor spin-exchange collisions, which could be a great advantage to avoid technical complications.

1. Introduction

Study on the production of spin-polarized species has been an important research subject for more than a few decades, since spin-polarized species are very useful tools to investigate various spin-dependent phenomena. There are essentially three kinds of spins to polarize, spin of isolated electrons, spin of a valence electron of atoms/ions, and spin of nucleus. In this paper, we theoretically discuss three different schemes to produce spin-polarized electrons and electron-spin polarized ions upon photoionization, some of which are accompanied by experimental results.

2. Spin-polarized electrons

Since photoelectric effect of a GaAs crystal has been found to be the most promising method as a spin-polarized source, lots of efforts have been made to improve the spin-polarization by fabricating a super lattice in a GaAs crystal, etc [1]. If rare gases can be used to produce spin-polarized electrons, instead, there are some advantages. Most importantly, it is free from not only a material damage but also periodic maintenance since it is a gaseous medium. The question we would like to address is, how much spin-polarization and electron yield one can achieve with rare gas atoms.

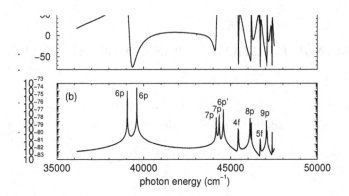

Figure 1. Three-photon ionization of Xe. (a) spin-polarization $P^{(3)}$ and (b) generalized ionization cross-section $\sigma^{(3)}$ as a function of photon energy. $6p$ etc. indicate two-photon intermediate resonances.

2.1. Photoionization of Xe

To answer this question, we have carried out theoretical calculations for Xe using multichannel quantum defect theory (MQDT) [2,3]. We have found that spin-polarization strongly varies as a function of laser wavelength, and the maximum spin-polarization is > 95% for all single-, two-, and three-photon ionization, if the wavelength is appropriately chosen [3].

Figure 1 shows specific results for three-photon ionization in terms of spin-polarization (Fig. 1(a)) and the generalized ionization cross section (Fig. 1(b)). Around photon energy of 38950 cm^{-1} ($\hbar\omega \simeq 4.8$ eV) the spin-polarization attains the value of $\sim$ 100 %. As shown in Fig. 1(b), the corresponding three-photon ionization generalized cross-section, however, has the rather low value of $\sim 10^{-80}$ cm^{6}sec^2. Nevertheless, this three-photon case might be of great interest to experimentalists since this photon energy corresponds to the third harmonic of well-developed Ti:Sapphire lasers. If the photon energy is 4.8 eV with the pulse energy of 1 mJ and the pulse duration of 200 fs, the intensity can be 5×10^{12} W/cm^2 if focused into the diameter of 350 μm. Given the generalized three-photon cross section of $\sim 10^{-80}$ cm^{6}sec^2, about 42% of the atoms in the interaction region is ionized. Therefore, if the Xe gas pressure is 1 Torr and the confocal parameter of the focusing (i.e., interaction length) is 1 cm, the electron yield would be 1.4×10^{13}/pulse. Of course, with higher gas pressure and/or more pulse energy, much more electron yield would be obtained.

3. Simultaneous production of spin-polarized electrons and electron-spin polarized ions

Spin-polarized electrons and electron-spin polarized ions can be complementary to each other, because their masses are different by three orders of magnitude. Indeed, the latter is a useful tool in surface physics [4,5] as well as in atomic/molecular physics. If spin-polarized electrons and ions are produced simultaneously, that can serve as a novel *dual source*.

Alkaline-earth (two-valence-electron) atoms are convenient for this purpose, since the photoion has only one valence electron whose degree of spin-polarization can be easily analyzed from the laser induced fluorescence [6]. After some theoretical analysis [7], we have found that spin-polarization of photoelectron and its counterpart, photoion is identical under certain conditions, leading to the simultaneous production of spin-polarized electrons/ions upon photoionization. In the following subsections we will consider two different schemes which utilize ns and ultrashort laser pulses.

3.1. Brute force method with ns pulses

First, we consider single-photon resonant two-photon ionization of alkaline-earth atoms [6], Sr, as shown in Fig. 2(a). Sr atoms in the ground $5s^2$ state are excited to the $5s5p\ ^3P_1$ state by the right-circularly polarized pump laser (689 nm), from which ionization takes place by absorbing photons from the linearly polarized ionization laser (308 nm). The essence of the scheme is that a *triplet state* must be involved in one way or another, since a singlet state does not have spin-orbit interactions that are necessary to transfer the angular momentum of absorbed photons into the spin angular momentum. That is, if $5s5p\ ^1P_1$ state is employed instead of $5s5p\ ^3P_1$, spin-polarization

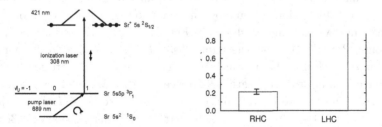

Figure 2. (a)Level scheme. (b)LIF intensity probed by the right-/left-circularly polarized probe laser, corresponding to the Sr$^+$ $5p\ ^2P_{1/2}$-$5s\ ^2S_{1/2}$ transition.

does not happen even if the pump laser is right-circularly polarized. Thus produced Sr$^+$ ions are in the ground $5s\ ^2S_{1/2}$ state (see Fig. 2(a)). In our experiment the intensities of the pump and ionization lasers are 44 kW/cm^2 and 2.5 MW/cm^2, respectively, resulting in the photoionization of ~25% of atoms in the interaction region. The degree of electron-spin polarization of Sr$^+$ $5s\ ^2S_{1/2}$ ions is experimentally determined from the ratio of the intensities of laser-induced fluorescence (LIF) by the right-/left-circularly polarized probe laser at 421 nm corresponding to the Sr$^+$ $5s\ ^2S_{1/2}$-$5p\ ^2P_{1/2}$ transition, i.e., $P = [I_{LHC} - I_{RHC}] / [I_{LHC} + I_{RHC}]$, where I_{RHC} and I_{LHC} being the LIF intensities by the right-/left-circularly polarized probe laser, respectively. Because of the geometry of the laser beams and the laser polarization we have chosen, spatial distribution of the LIF emission is uniform around the probe beam axis. Clearly, no calibration for the detection efficiency of I_{RHC} and I_{LHC} is needed. Note that all laser beams described above are in the cross-beam geometry [6] at right angle so that the transitions illustrated in Fig. 2(a) take place.

Figure 2(b) shows the experimentally measured LIF intensity by the right-/left-circularly polarized probe laser. From the data, the degree of spin-polarization is found to be 64±9% [6], which agrees well with the theoretical prediction [7]. If different polarization is used for the excitation/ionization lasers, different degree of spin-polarization would be obtained. For the particular case with the ionization laser being vertically/horizontally linear polarization, we have recently obtained experimental results which agree well with our theoretical prediction [8].

3.2. *Pump-probe method with ultrafast pulses*

In the previous subsection, it is ns pulses that are used to produce spin-polarized electrons/ions from two-valence-electron atoms, in particular, Sr. If ultrafast laser pulses are used, instead, an interesting scenario emerges [9,10]. For illustration we employ another two-valence-electron atom, Mg, with the level scheme as shown in Fig. 3. We excite Mg atoms in the ground $3s^2\ ^1S_0$ state to the triplet $3s3p^3P_1(M = +1)$ state by a right-circularly polarized laser, which will serve as an initial state for the subsequent pump-probe ionization [10]. Once $3s3p^3P_1(M = +1)$ is populated, we send an ultrashort pump pulse to coherently excite $3s3d\ ^3D_{1,2}$. Therefore, the excited state wavefunction, $|\Psi(t)\rangle$, is a superposition of $3s3d\ ^3D_1$ and $^3D_2$, i.e., $|\Psi(t)\rangle = (-1/2\sqrt{6})|3s3d\ ^3D_1\rangle + (\sqrt{3}/2\sqrt{10})|3s3d\ ^3D_2\rangle e^{-i\Delta t}$, where the time origin of t is chosen to be the instant of the pump pulse, and Δ is an energy difference between $3s3d$ $^3D_1$ and $^3D_2$. Note that state-flipping takes place after the pump pulse due to the exponential factor, $e^{-i\Delta t}$.

Since each of $|3s3d\ ^3D_1\rangle$ and $|3s3d\ ^3D_2\rangle$ can be further decomposed into a superposition of spin-up/-down electron wavefunctions of two valence electrons, state-flipping can be interpreted as spin-flipping. In order for this scenario to work, it is essential that the pulse durations of the pump and probe pulses are much

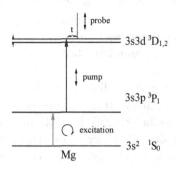

Figure 3. Level scheme for the ultrafast spin-polarization.

shorter than the period given by Δ^{-1} so that state-flipping does not occur during the pump and probe pulses. For this specific example of Mg, the pulse durations of the pump and probe pulses can be ps, since the fine structure splitting of $3s3d\ ^3D_{1,2}$ is 0.031 cm^{-1} corresponding to the 1.1 ns fine structure coupling time. Naturally, the fine structure splitting is larger for heavier atoms or ions, and a shorter pulse is needed. After a certain delay time, the probe pulse is turned on to ionize atoms in the coherently excited state, $\Psi(t)$. By using Eq. (3) in Ref. [10], it is straightforward to compute the photoelectron/photoion yields with spin-up/-down.

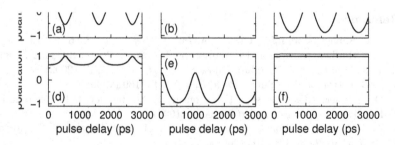

Figure 4. Spin-polarization as a function of pulse delay. (a) ω_{probe}=4.01 eV with linear polarization, (b) ω_{probe}=4.01 eV with left-circular polarization, (c) ω_{probe}=4.01 eV with right-circular polarization, (d) ω_{probe}=4.46 eV with linear polarization, (e) ω_{probe}=4.46 eV with left-circular polarization, and (f) ω_{probe}=4.46 eV with right-circular polarization.

Figure 4 shows the computed spin-polarization as a function of pulse delay for two different photon energies, ω_{probe}=4.01 eV and 4.46 eV, with three different polarization (linearly or left-/right-circularly polarized) for the probe laser. Note that the polarization of the excitation laser and the pump laser have been fixed to be right-circular and linear, respectively, for all Figs. 4(a)-(f). Clearly the behavior of spin-polarization is very different for different photon energies and polarization of the probe laser. Most interestingly, as shown in Fig. 4(a), even the sign of spin-polarization may be controlled simply by changing the pulse delay. We note that the change of the sign in spin-polarization with a pulse delay cannot happen for a single-valence-electron atom [9].

4. Summary

In summary, we have discussed three different schemes to polarize spin of electrons and electron-spin of ions by ns and ultrashort (ps) laser pulses. Our schemes are direct method, and involve neither optical pumping nor spin-exchange collisions. If the production of the electron/ion yield may be in pulsed operation, use of pulsed lasers can be a great advantage compared with CW lasers because of the much higher energy (more than several orders of magnitude) available per unit time and accessibility to the wide range of wavelength.

5. Acknowledgments

This work is supported by the Grant-in-Aid for scientific research from the Ministry of Education and Science of Japan.

References

1. A.V. Subashiev and J.E. Clendenin, *Int. J. Mod. Phys. A* **15**, 2519 (2000).
2. A. L'Huillier, X. Tang and P. Lambropoulos, *Phys. Rev. A* **39**, 1112 (1989).
3. T. Nakajima and P. Lambropoulos, *Europhys. Lett.* **57**, 25 (2002).
4. S. Grafström and D. Suter, *Phys. Rev. A* **54**, 2169 (1996).
5. D.L. Bixler, J.C. Lancaster, F.J. Kontur, P. Nordlander, G.K. Walters, and F.B. Dunning, *Phys. Rev. B* **60**, 9082 (1999).
6. T. Nakajima, N. Yonekura, Y. Matsuo, T. Kobayashi, and Y. Fukuyama, *Appl. Phys. Lett.* **83**, 2103 (2003).
7. T. Nakajima and N. Yonekura, *J. Chem. Phys.* **117**, 2112 (2002).
8. Y. Matsuo, T. Kobayashi, N. Yonekura, and T. Nakajima (to be submitted).
9. M.A. Bouchene, S. Zamith, and B. Girard, *J. Phys. B* **34**, 1497 (2001).
10. Takashi Nakajima, *Appl. Phys. Lett.* **84**, 3786 (2004).

AN ATTEMPT TOWARD DYNAMIC NUCLEAR POLARISATION FOR LIQUID ^{3}HE

T.IWATA, S.KATO, H.KATO, T.MICHIGAMI, T.NOMURA, T.SHISHIDO,
Y.TAJIMA, H.UENO, H.Y.YOSHIDA
Yamagata University,
1-4-12, Kojirakawa-cho, Yamagata,
990-8560, JAPAN
E-mail: tiwata@sci.kj.yamagata-u.ac.jp

Dynamic nuclear polarisation of liquid ^{3}He is pursued in zeolite doped with TEMPO free radical. The zeolite was successfully doped with TEMPO. The ESR signal of TEMPO inside the zeolite showed the free radical molecules are dispersed at some level. Good Stability of TEMPO trapped in zeolite was confirmed in ESR measurements.

1. Introduction

Polarised ^{3}He targets have been employed as a polarised neutron target. Since only neutron inside ^{3}He is polarised, they are good targets for the study of neutron itself in scattering experiments. They have been realized by optical pumping technique. However, as it is applicable only for gas, the density of the targets are limited. And their application is also limited.

If ^{3}He is polarised in dense form, i.e. in liquid or solid, it will open a door to various applications. Some attempts have been made in this direction. One of them is made by brute force method. Rather high polarisation was obtained in solid phase at high magnetic field, at a low temperature and a high pressure[1]. Polarised liquid was also obtained by quickly melting the polarised solid [2]. However, its application is limited due to such an extreme condition.

Alternative way is Dynamic Nuclear Polarisation(DNP). DNP by direct coupling from the electron system to ^{3}He was proposed by Delheiji et al. [3]. They used diluted paramagnetic centers in solid ^{3}He. However, they obtained no polarisation enhancement. Another way is based on the coupling between ^{3}He and polarised material with large surface area. Nuclei in the material are polarised by DNP and their polarisation is transferred to the surrounding ^{3}He. The idea was tested with naturally existing paramagnetic centers in Teflon [4] and charcoal [5]. Small enhancements were observed. Most recently, PSI group made systematic study with polyethylene, Teflon, and zeolite doped with TEMPO free radical [6]. No enhancement was observed in polarisation, although they observed change of the NMR signal of ^{3}He. In their study, doping of free radical was not successful in particular for Teflon and Zeolite.

2. Direct coupling in Zeolite

Our idea is based on the direct coupling between a free radical and ^{3}He. It is similar to that of PSI group in the point that we also use a material doped with a free radical. Firstly, we embed a free radical into a porous material. Secondly, it is filled with liquid ^{3}He. Finally, coupling between the free radical and the ^{3}He is induced by microwave. One of the key issues is embedding free radical molecules. It is necessary that they should be firmly trapped in the material. In addition, they should be well dispersed. For these purposes, matching the cavity size to the free radical molecule is important. We chose NaY type zeolite as shown in FIgure 1, with a combination of the TEMPO free radical.

2.1. zeolite

Zeolite is a compound basically made of Aluminum, silicon and oxygen. The atoms are networked to each other and give regularly ordered cavity structure. The NaY type zeolite has a cavity with a diameter of 13 Å.
The cavity, called super cage, has a window of which diameter is 7.4 Å. The molecule of TEMPO free radical which has the size of 6 to 8 Å, just fits the super cage.

2.2. Doping process

The doping process of TEMPO is following. In advance, the zeolite powder is activated at a temperature of 500 °C for 8 hours. The TEMPO is dissolved into n-pentane. The solution stored in a closed bottle is stirred with the zeolite powder for 8 hours. In the open air, the n-pentane evaporates away. This method is applied for studies of unstable radicals in the field of chemistry. The amounts of the TEMPO and the zeolite were tuned to give spin density of the order of $10^{19} spins/cc$ in zeolite. The amounts correspond to 16 % of the super cages occupied with the TEMPO molecules.

Figure 1. Zeolite unit cell. The circles represent Silicon or Aluminum atoms. The lines represent Oxygen atoms.

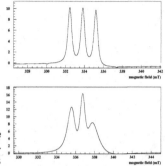

Figure 2. ESR signal of TEMPO trapped in zeolite(lower side) and dissolved in ethanol.

2.3. ESR signals

The ESR signal of TEMPO in zeolite is is compared with the ESR signal of TEMPO dissolved by ethanol as shown in Figure 2. Usually, in solution very narrow signals are obtained, because the molecules of solution are moving around and TEMPO molecules are well dispersed. Comparing with such a narrow signal, the signal in zeolite is a little broader. However, the peaks are still separated, which is a feature of TEMPO. This means the TEMPO molecules in zeolite are dispersed at some level.

2.4. Stability of TEMPO

Another important issue is the stability of the TEMPO in zeolite. The intensity variation of the ESR signal was measured. Figure 3 (left figure) shows the intensity variation of the ESR signal of TEMPO in zeolite in the open air at room temperature. Although the intensity suddenly dropped by about 20 %, it stayed at the same level for more than 100 hours. The measurement was also carried out with the zeolite in vacuum. Cconstant intensity was observed again for a long period as shown in Figure 3 (right figure).

This means the TEMPO molecules are firmly trapped in zeolite.

For comparison, the same measurement was made for a polyethylene sheet doped with TEMPO by diffusion. The intensity decreased with a time constant of about 5 hours. This is for the following reasons: The TEMPO molecules are trapped only in the amorphous part of polyethylene. At room temperature the polyethylene molecules are still moving. The

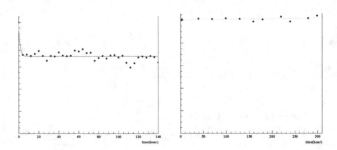

Figure 3. Intensity variation of ESR signal when the zeolite doped with TEMPO is in the open air at room temperature(left) and in vacuum at room temperature(right).

movement enhances the diffusion of TEMPO.

3. Conclusions

DNP for liquid ^{3}He is pursued through the direct interaction between ^{3}He and a free radical molecule embedded in the cavities of a zeolite. We have prepared the zelite doped with TEMPO free radical. Being well dispersed, the TEMPO molecules are firmly trapped in zeolite. We are ready to make DNP for liquid ^{3}He in the zeolite.

Acknowledgments

This work was supported by Grants-in-Aid for Scientific Research (Exploratory Research, No.16654036) of MEXT, Japan.

References

1. G.Bonfait et al., *Phys.Rev.Lett.* **53**, (1984) 1092
2. M.Chapellier, G.Frossati, and F.Rasmussen, *Phys.Rev.Lett.* **42**, (1979) 902
3. P.P.J.Delheij, H.Postma and K.Bindels, International Symp. High Energy Spin Physics, Bonn, 1990, eds, W.Meyer et al., Springer Verlage Berlin 1991, vol.2, p.321
4. A.Shuhl et al., *Phys.Rev.Lett.* **54**, (1985) 1952
5. L.W.Engel and Keith DeConde, *Phys.Rev.* **33**, (1986) 2035
6. B.van den Brandt, E.I.Bunyatova, P.Hautle, J.A.Konter, A.I.Kovalev, S.Mango, Nuclear Instruments and Methods in Physics Research Section A: Accelerators, Spectrometers, Detectors and Associated Equipment, Volume 356, Issue 1, 1 March 1995, Pages 138-141

Poster Presentations

PERFORMANCE OF GAASP/GAAS SUPERLATTICE PHOTOCATHODES IN HIGH ENERGY EXPERIMENTS USING POLARIZED ELECTRONS[*]

A. BRACHMANN, J.E. CLENDENIN, T. MARUYAMA, E.L. GARWIN, K. IOAKEIMIDI,
C.Y. PRESCOTT, J.L. TURNER

*Stanford Linear Accelerator Center, 2575 Sand Hill Road,
Menlo Park, CA 94025, USA*

R. PREPOST

*Department of Physics, University of Wisconsin
Madison, WI 53706, USA*

The GaAsP/GaAs strained superlattice photocathode structure has proven to be a significant advance for polarized electron sources operating with high peak currents per microbunch and relatively low duty factor. This is the characteristic type of operation for SLAC and is also planned for the ILC. This superlattice structure was studied at SLAC [1], and an optimum variation was chosen for the final stage of E-158, a high-energy parity violating experiment at SLAC. Following E-158, the polarized source was maintained on standby with the cathode being re-cesiated about once a week while a thermionic gun, which is installed in parallel with the polarized gun, supplied the linac electron beams. However, in the summer of 2005, while the thermionic gun was disabled, the polarized electron source was again used to provide electron beams for the linac. The performance of the photocathode 24 months after its only activation is described and factors making this possible are discussed.

1. Photocathode History and Operation

The cathode was installed in the preparation chamber in June 2003, heat-cleaned once at restricted high temperatures, activated with Cs and NF_3, inserted into the

---

[*] This work is supported by Department of Energy contracts DE-AC02-76SF00515 (SLAC) and DE-AC02-76ER00881 (UW).

gun under vacuum, and thereafter only Cs was added periodically and *in situ*. This cathode provided a highly polarized (~90%), high intensity (~5×10^{11} e⁻ per square-shaped macropulse of ~300 ns) pulse during a dedicated 2-month run at 120 Hz. In the summer of 2005, the cathode was used to support PEPII operations 2005 (30 Hz, $2*10^{10}$ e⁻ per pulse, 2 ns pulse). At the end of this period the cathode was in use for more than 900 days, while it delivered electrons for the accelerator for almost 200 days. During the entire time, about 160 (automated) cesiations have been performed in 5 to 6 day intervals. RGA measurements near the gun indicate a total residual gas pressure of 10^{-11} Torr. Excluding H_2, the vacuum near the cathode is less than 10^{-12} Torr (Figure 1). Quantum efficiency (QE) profile measurements of the 2-cm-diameter emitting area have been performed several times since installation of the cathode. During this time the QE profile showed a small decrease near the center of the cathode, which can be attributed to 'back–bombardment' of the cathode's surface by residual ions. Our experience shows that recovery of the original QE profile by re-cesiation is only partly possible. However, for our operating conditions this is a negligible phenomenon as the QE monitor indicates stable performance throughout this time (Figure 2). Due to operational constraints, polarization measurements could not be performed after September 2003. Nevertheless, our results show that GaAsP/GaAs superlattice photocathodes can be low–maintenance, highly reliable e⁻ sources for future high energy physics projects such as the International Linear Collider.

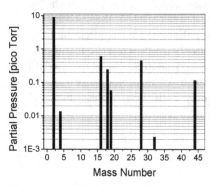
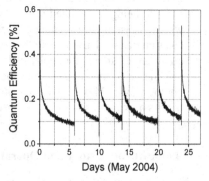

Figure 1: Residual Gas Analysis near Photocathode.

Figure 2: Quantum efficiency monitor showing cesiation cycle of 5 – 6 days.

References

1. T. Maruyama et al., *Appl. Phys. Lett.* **85**, 2640 (2004).

DEVELOPMENT OF SPIN-EXCHANGE TYPE POLARIZED ^{3}HE TARGET FOR RI-BEAM EXPERIMENTS

K. ITOH

Department of Physics, Saitama University,
Shimo-Okubo 255,
Sakura, Saitama 338-8570, Japan
E-mail: keisuke@ne.phy.saitama-u.ac.jp

A spin-exchange type polarized ^{3}He target for RI beam experiments have been developing at CNS. When we apply the target to RI-beam experiments, one of big problems is the thick cell material. Thus, we recently developed target cells that partly used thin titanium foils instead of glass. The resulting spin relaxation time of the "Ti-cell" was 3.7 hours.

1. Polarized ^{3}He target for RI beam experiments

A spin-exchange type polarized ^{3}He target for RI beam experiments have been developing at CNS. In the spin-exchange method, the ^{3}He gas in a glass cell are polarized via the spin-exchange reaction with Rb atoms. We employ this method because it enables us to make a high density ^{3}He target rather easily. ¿From the viewpoint of the count rate, high density of the target is suitable for RI-beam experiments.

When we apply the target to RI-beam experiments, one of big problems is the thick cell material. It is not only causing of background events, but also makes difficult to detect low energy recoil particles. As a result, thick cell materials makes reaction channel identification difficult. Therefore, we started to develop the target cells that partly used thin metal foils.

2. Development of new target cell that partly used thin metal foils

2.1. Choice of foil material

There are some requirement of the foil materials. Firstly, foil materials must be tolerant to nitric acid. Because, in order to remove paramagnetic impurities, the inner wall of the target cell must be cleaned with nitric acid.

Secondly, foil materials must not be broken by the pressure of several atoms. Because, required pressure of ^{3}He gas in a cell is about 3-8 atm. Thirdly, spin relaxation time is long. We choice Ti foils for the foil materials by the first and second requirement.

2.2. Cell demension

We prepare the cell which partly used Ti foils and measure the relaxation time. The glass cell has a so-called "double cell" structure and consists of a target cell and an optically pumping cell, connected to each other through a pipe with an inner diameter of 10 mm. The target cell has cylindrical shape (ϕ 50 mm × 100 mm). The entrance and exit window which pass through RI beams are covered by 50 μm^t Ti foils instead of glass. The foils are glued to the glass by using STYCAST 1266. The pumping cell also has cylindrical shape (ϕ 60 mm × 70 mm). The volume of target cell and pumping cell are both 200 cm^3.

2.3. Cold relaxation time measurement

In the cold relaxation process, in which the cell temperature is room temperature, the polarization depends only the spin relaxation rate of ^{3}He. Time dependence of polarization was measured by the AFP-NMR technique and the spin relaxation time of the cell was deduced. Measured cold relaxation time for the "Ti-cell" was $\sim$ 3.7 hours. Maximum ^{3}He polarization in this cell was expected to be $\sim$ 10 %.

3. Conclusion and Perspective

Measured relaxation time of the "Ti-cell" was 4-5 times shorter than normal cell which is made from glass only. Polarization of 10% is small for ($^3\vec{\text{He}}, \alpha$) reaction measurement with RI beam, which we plan. Search and test for other materials are needed. Demagnetization of the target cell is also needed.

PRODUCTION OF SPIN-POLARIZED RI BEAMS VIA PROJECTILE FRAGMENTATION AND THE APPLICATION TO NUCLEAR MOMENT MEASUREMENTS

D. KAMEDA, H. UENO, K. ASAHI, A. YOSHIMI, H. WATANABE,
T. HASEYAMA, AND Y. KOBAYASHI
RIKEN, 2-1 Hirosawa, Wako, Saitama 351-0198, Japan
E-mail: kameda@rarfaxp.riken.jp

M. UCHIDA, H. MIYOSHI, K. SHIMADA. G. KIJIMA, M. TAKEMURA,
D. NAGAE, G. KATO, S. EMORI, S. OSHIMA, T. ARAI AND M. TSUKUI
Department of Physics, Tokyo Institute of Technology,
2-12-1 Oh-okayama, Meguro-ku, Tokyo 152-8551, Japan

Spin-polarized radioactive ion beams of ^{30}Al and ^{32}Al were produced via projectile fragmentation using 95 MeV/nucleon ^{40}Ar projectiles. The static magnetic moments for the both isotopes and the static electric quadrupole moment for ^{32}Al were measured by means of the β-NMR technique taking advantage of the polarizations around a percent.

In the mass $A \lesssim 40$ region, various spin-polarized unstable nuclei far from stability have been obtained from projectile-fragmentation (PF) reactions to study the electromagnetic moments. Through the studies, we have observed a trend of the polarization phenomena in the PF reactions that the magnitudes of the polarizations become smaller if the difference of masses between the unstable nucleus (fragment) and the projectile becomes larger. For example, the polarizations of the fragments with 2-nucleon removals from projectiles could be obtained around ten percents[1,2]. In the case of 5-nucleon removals, the polarizations were observed to be only a few percents[3]. To study nuclear moments in the heavier mass regions far from stability, we need to confirm whether or not the sufficient polarizations can be obtained for fragments with much larger removed-nucleon numbers.

Recently, we have produced polarized ^{30}Al and ^{32}Al beams from the fragmentation of 95 MeV/nucleon ^{40}Ar projectiles, involving 10 and 8 removed-nucleon numbers, respectively. The magnetic (μ) moments for the both isotopes and the electric quadrupole (Q) moment for ^{32}Al have been measured by means of the β-NMR technique taking advantage of the sufficient polarizations around 1 %. The polarized unstable nuclei were obtained using RIKEN Projectile-fragment separator (RIPS)[4]. The ^{40}Ar projectiles were bombarded to Nb targets with the thicknesses of 0.13 g/cm^2 and 0.37 g/cm^2 for ^{30}Al and ^{32}Al, respectively. To produce polarizations in the fragments emitted from the targets, the emission angles within 1.3° − 5.7° were accepted by RIPS. The outgoing momenta of the ^{30}Al (^{32}Al) fragments were selected to the range of 12.4 − 12.7 GeV/c (12.2 − 13.0 GeV/c) using RIPS. The range corresponded to 1.006 − 1.026p_0 (0.975 − 1.034p_0), where p_0 denotes the momentum at the peak of the momentum distribution for ^{30}Al (^{32}Al). The polarized beams were implanted in a single crystal α-Al$_2$O$_3$ to which a static magnetic field $\sim$ 500 mT was applied to perform the β-NMR experiments. The detail is described in Ref. 5, 6.

The present work has provided a clear prospect for studying nuclear moments in the wide mass regions reached via PF reactions that spin polarizations useful in the β-NMR experiments can be obtained for fragments produced via many-nucleon removals as large as 10.

Acknowledgments

The authors wish to thank staffs of the RIKEN Ring Cyclotron for their support. The authors (D.K. and T.H.) are grateful for the Special Postdoctoral Researcher Program in RIKEN. This work was supported in part by a Grant-in-Aid for Scientific Research from the Ministry of Education, Science, Sports and Culture.

References

1. K. Asahi, et. al., *Phys. Lett.* **B251** 488 (1990).
2. H. Okuno, et. al., *Phys. Lett.* **B335** 29 (1994).
3. H. Ueno, et. al., *Phys. Rev.* **C53** 2142 (1996).
4. T. Kubo, et. al., *Nucl. Instr. and Meth.* **B70** 309 (1992).
5. H. Ueno, et. al., *Phys. Lett.* **B615** 186 (2005).
6. D. Kameda et. al., *RIKEN Accel. Prog. Rep.* **38** (2005) in press.
7. B. H. Wildenthal, *Prog. Part. Nucl. Phys.* **11** 5 (1984).

A NEW ^{3}HE POLARIZER AND TARGET SYSTEM FOR LOW-ENERGY SCATTERING MEASUREMENTS *

T. KATABUCHI, T. B. CLEGG, T. V. DANIELS AND H. J. KARWOWSKI

Department of Physics and Astronomy, University of North Carolina,
Chapel Hill, North Carolina 27599-3255, USA
and Triangle Universities Nuclear Laboratory (TUNL),
Durham, North Carolina 27708-0308, USA
E-mail: katabuchi@tunl.duke.edu

We have developed a new polarized ^{3}He target system which facilitates p-^{3}He elastic scattering at proton energies below 5 MeV. This system consists of a target cell placed in a compact, shielded sine-theta coil inside a scattering chamber and an external optical pumping station utilizing Rb spin-exchange. Computer-controlled valves allow polarized ^{3}He gas to be transferred quickly between the optical pumping station and the target cell. Target gas polarimetry is accomplished using NMR and calibrated using the known analyzing power of α-^{3}He scattering.

A new polarized ^{3}He target system has been developed[1] with the initial experimental goal of measuring spin-dependent observables in the p-^{3}He elastic scattering at incident proton energies below 5 MeV. Microscopic calculations, based on modern nucleon-nucleon forces determined precisely from two-nucleon experimental data, are found to underpredict the proton analyzing power in p-^{3}He elastic scattering at 1.20 and 1.69MeV[2]. Comparison with further accurate experimental data for the p-^{3}He scattering, for different spin observables, are desired to reduce theoretical ambiguities[3].

Our polarized target system is designed to utilize spin-exchange optical pumping with rubidium to polarize ^{3}He, thereby allowing experiments at a high counting rate from a high target density. We briefly describe the polarized target system here. Further datails have been reported in Ref. 1. The system consists of two separate parts: a unique Pyrex target cell and a separate optical pumping station. They are connected through a computer-

---

*This work is supported in part by the U.S. Department of Energy, Office of High Energy and Nuclear Physics, under Grant No. DE-FG02-97ER41041

controlled valve manifolds to facilitate moving ^{3}He gas quickly when needed between the optical pumping cell and the target cell.

The optical pumping station is comprised of an optical pumping cell inside an isolated oven, a laser, optical components and a valved manifold. The pumping cell, oven and manifold are placed inside a 29.4-cm diameter solenoid coil shielded externally and on both ends by mu-metal, which provides a highly uniform internal B-field of ∼0.7 mT. The GE180 aluminosilicate glass optical pumping cell with 200-cm^3 inner volume was found to have a 36-h spin-relaxation time. Light from a fiber-coupled diode laser array passed through optical components which produced the circular polarization needed before this light impinged on Rb atoms in the optical pumping cell.The oven and optical pumping cell were heated by the ∼ 60 W of laser light and their temperature was held at ∼ 185°C by regulating the flow of cooling air to the oven's interior. Maximum ^{3}He polarizations of up to ∼ 30 % have been obtained.

The target cell is made of Pyrex glass and is roughly spherical with a diameter of 5 cm. The cell has beam entrance and exit apertures and windows on either side of the beam where scattered particles emerge. The apertures and windows are covered with 25 and 7.5 μm thick Kapton foil, respectively. A uniform B-field is provided over the target cell with a shielded, compact sine-theta coil. This consists of 24 straight, copper rods parallel to and spaced equally around the axis of a 7.5 cm diameter, 30 cm long cylinder, which itself is aligned with the beam axis. The rod currents are adjusted to be proportional to the sine of the azimuthal angle from the quantization axis established by the B-field. Typical target cell spin relaxation times ranging between 2 and 3 h are largely dictated by the presence of the epoxy and Kapton needed for windows. Refreshing polarized gas frequently in the target cell accommodates such short spin-relaxation times. The target polarization is monitored using pulsed NMR with a coil attached to the cell. The NMR polarimetry was calibrated by measuring the ^{3}He asymmetry in α-^{3}He elastic scattering at $E_\alpha = 15.3$ MeV and $\theta_{^3He} = 47°$ in laboratory, where the absolute value of the analyzing power A_y is 1.

References

1. T. Katabuchi et al., Rev. Sci. Instr. **76**, 033503 (2005).
2. M. Viviani et al., Phys. Rev. Lett. **86**, 3739 (2001).
3. B. M. Fisher et al., submitted to Physical Review C.

POLARIZATION DATA ANALYSIS OF THE COMPASS ^{6}LiD TARGET

J. KOIVUNIEMI,[*] N. DOSHITA, Y. KISSELEV, K. KONDO, W. MEYER, G. REICHERZ

Physics Department, University of Bochum, 44780 Bochum, Germany

F. GAUTHERON

Physics Department, University of Bielefeld, 33501 Bielefeld, Germany

The COMPASS experiment at CERN has been taking spin physics data in 2002 - 2004 with a double cell solid ^{6}LiD target with high +57 % and -48 % nuclear polarization. The measured asymmetries give access to the longitudinal spin structure function[1], transverse Collins and Sivers asymmetries[2], and to gluon polarization[3]. Stable and reproducible operation of the polarized target is essential in the typically 100 day long data taking periods. We discuss our present understanding of the polarization properties of the ^{6}LiD target measured with continuous wave nuclear magnetic resonance.

The COMPASS polarized deuteron target[4,5,6] has 60 cm long and 3 cm diameter target cells with 424 cm^3 cell volume. The cells are inside glass fiber mixing chamber with liquid ^{3}He/^{4}He mixture at 50 - 300 mK temperature with four NMR coils on the upstream target cell and four on the downstream cell. Two halves of the microwave cavity are separated with a microwave stopper[4]. The produced particles inside acceptance of 69 mrad are detected in the spectrometer. The 2 - 4 mm ^{6}LiD crystals[7] have face-centered cubic structure and they were irradiated at Bonn with 20 MeV electron beam to create paramagnetic centers into the material[8].

During the initial polarization calibration around 1 K in 2.5 T field[9] the spin system reaches thermal equilibrium (TE) with helium in about 15 hours. Primary thermometer based on ^{3}He vapor pressure given by the International Temperature Scale ITS-90[10] is used. The error in the determination of the calibration spin temperature is typically 0.3 - 0.7 %. Monte

---
[*]Email: Jaakko.Koivuniemi@cern.ch

Carlo simulations of the TE-calibration data analysis indicate an error of 0.3 - 0.5 % for the Curie constants with the 5 - 12 day long calibrations.

During the physics data taking the positive polarization was +57 % and negative -48 %, but also symmetric cell polarization +53 % and -53 % was reached. The polarization measured by different coils is compared to the average cell polarization in Fig. 1. After the start of each polarization build up a stable distribution is reached fast with ±6 % from the cell average. The distribution is quite stable for the whole run. It is not very sensitive to changes in microwave field or in ^{3}He flow in the dilution cryostat. The

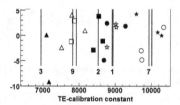 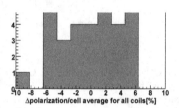

Figure 1. *Left:* Thermal equilibrium calibration constants vs. relative polarization of coils 1 - 4 and 7 - 10 for runs 2002 - 2004. No correlation between coil position and polarization is seen. There is a symmetry in the calibration constants with respect to the microwave stopper in the center. *Right:* Histogram of relative coil polarizations 2001 - 2004 for four upstream and four downstream coils. In the technical run of 2001 the coils were embedded into the target material. The deviation in the measured polarization seems to be less than 6 % relative.

polarization is lost about 0.05 %/day in frozen spin mode in 2.5 T field below 90 mK and 0.7 %/day in 0.42 T during transverse data taking.

References

1. E.S. Ageev et al., Phys. Lett. B **612**, 154 (2005).
2. V.Yu. Alexakhin et al., Phys. Rev. Lett. **94**, 202002 (2005).
3. E.S. Ageev et al., Phys. Lett. B **633**, 25 (2006).
4. J. Ball et al., Nucl. Instr. Meth. **A498**, 101 (2003).
5. N. Takabayashi, PhD theses, Nagoya University (2002).
6. D. Adams et al., *Nucl. Instr. Meth.* **A437**, 23 (1999).
7. S. Neliba et al., *Nucl. Instr. Meth.* **A526**, 144 (2004).
8. A. Meier, PhD theses, University of Bochum (2001).
9. K. Kondo et al., *Nucl. Instr. Meth.* **A526**, 70 (2004).
10. H. Preston-Thomas, *Metrologia* **27**, 3 (1990).

LAMB-SHIFT POLARIMETER FOR DEUTERON GAS TARGET AT LNS

ITARU NISHIKAWA AND TADAAKI TAMAE
Laboratory of Nuclear Science, Tohoku University,
1-2-1,Mikamine Sendai 982-0826, Japan

A Lamb-shift type polarimeter for polarized deuteron gas target was constructed and the test operation was done by using an unpolarized deuterium ion source.

1. Spin-Filter Polarimeter at LNS

We have developed a Lamb-shift type polarimeter for an optical pumping deuteron gas target. The polarimeter is conventional one and composed of a d-Cs charge-exchange section, the spin-filter section and the Stark-Quenching & Lyman-α photon counting section. A cross-sectional view of the polarimeter is shown in Fig.1. The Cs vapor cell is an assembly of copper sleeves and the outer ceramic sleeves warmed by nichrome ribbon heater. The spin-filter cavity is made of brass and split into 4-quadrant parts. Two of the quadrants are electrically biased to make a transverse electric field on a beam axis. At the cavity central region a quasi-uniform magnetic field is generated by three main coils and two sub coil sets. In the last stage of the polarimeter the nuclear-spin-filtered D(2S) atoms are forcedly Stark-quenched at front of the cylindrical electrode and Lyman-α decay photons are detected by an AC-coupled photo-multiplier tube.

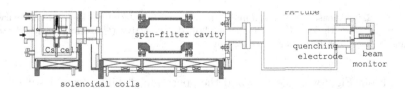

Figure 1. A view of the spin-filter polarimeter I :Charge-Exchange section II :Spin-Filter section III :Stark-Quenching & Photon Counting Section

2. Test Operation of the Polarimeter

The test operation of the polarimeter was done by using unpolarized deuterons from a test RF deuterium ion-source of a Pyrex tube with a RF coil. The deuterium ions extracted from the ion-source were accelerated to 1.5 keV, mass-analyzed with a permanent magnet Wien-filter and then partially converted into D(2S) atoms in the Cs cell of the polarimeter. The monitored deuteron beam current at the end of the polarimeter was typically 0.5nA. The relative transmission of the D(2S) through the spin-filter section was measured with sweeping the magnetic field. A result is shown in Fig.2. The three peaks in the data reflects the unpolarized population of magnetic sub-state of the deuteron from the test ion-source. In Fig.2 the vertical axis is the averaged photon count rate in 10 sec and the horizontal axis is the magnetic flux averaged in the cavity central region.

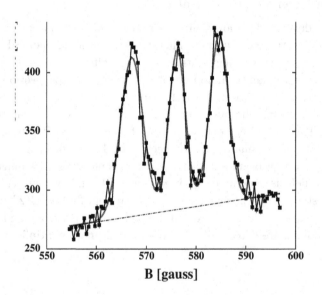

Figure 2. A measured deuteron magnetic sub-state population peaks and fitting curve, solid square with solid line: data point, dashed line: fitting (Gaussian+linear), dotted and dashed line: linear part of the fitting

From Fig.2 we note S/N and the operation stability are not sufficient, the suppression of background photons and the improvement of the uniformity of the magnetic field are required. We are also developing an ion extraction system using an e-gun for the polarimetry of the polarized target.

PERFORMANCE EVALUATION OF NPOL AT RIKEN

S. NOJI, K. MIKI, K. YAKO, H. SAKAI, AND H. KUBOKI
Department of Physics, University of Tokyo, Bunkyo, Tokyo 113-0033, Japan
E-mail: noji@nucl.phys.s.u-tokyo.ac.jp

T. KAWABATA AND K. SUDA
Center for Nuclear Study, University of Tokyo, Bunkyo, Tokyo 113-0033, Japan

K. SEKIGUCHI
Institute of Physical and Chemical Research (RIKEN),
Wako, Saitama 351-0198, Japan

We constructed a neutron polarimeter NPOL at RIKEN to test Bell's inequality in a proton-neutron system. NPOL consists of 12 planes of two-dimensional position-sensitive plastic scintillators. Azimuthal distributions of the $n + p$ scattering in NPOL yield neutron polarization. We evaluated the performance of NPOL with the polarized neutrons from the ^{6}Li($\vec{d}, \vec{n}$)X reaction at $E = 135A$ MeV and obtained an effective analyzing power of 0.28 and a figure of merit of 3.2×10^{-5}.

We constructed a neutron polarimeter NPOL at RIKEN to measure a spin correlation between protons and neutrons in $^1S_0$ state produced by the $d(d, pn)pn$ reaction at $E = 135A$ MeV for a test of Bell's inequality.

Figure 1 shows a schematic view of NPOL. NPOL is a set of 12 planes of neutron detectors with a size of $60 \times 60 \times 3$ cm^3, each of which consists of six plastic scintillator bars (BC408). On both ends of the bars PMTs (H7195 or H7415) are glued. Timing information of PMTs allows us to deduce interaction positions of neutrons in NPOL. Typical position resolution is ± 3 cm in the longitudinal direction of the bar. The position information can be converted into the scattering angles θ and ϕ. NPOL measures neutron polarization using one plane as a neutron scatterer and another as a neutron catcher. The neutron polarization is the scattering asymmetry in NPOL normalized by its effective analyzing power A_y^{eff}. A figure of merit (FOM) is defined by $\epsilon_{\text{double}} (A_y^{\text{eff}})^2$, where ϵ_{double} is the double scattering efficiency. The most useful event to get a large FOM value is the elastic scattering of a neutron by a proton in a scatterer and detection of the scattered neutron in a catcher.

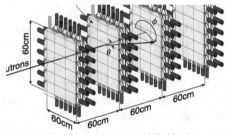
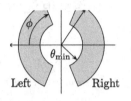

Figure 1. A schematic view of NPOL. Scattering angles θ and ϕ are indicated. Thin scintillators (CP VETO) in front of each stack is used to tag charged particle events.

Figure 2. The sector definition for left/right asymmetry.

The measurement of A_y^{eff} and ϵ_{double} was performed with the polarized neutrons from the $^6\text{Li}(\vec{d},\vec{n})\text{X}$ reaction at $0°$ at $T_d = 270\,\text{MeV}$. Neutron energy was $\sim 135\,\text{MeV}$, and its flux was $1 \times 10^9/\text{s}$ in NPOL located 18 m downstream of the target. The deuteron polarization p_d was ± 0.64. Assuming the polarization transfer coefficient to be $K_y^y(0°) = 2/3$ [2], we estimated the neutron polarization to be $p_n = -\frac{3}{2}K_y^y p_d \simeq \mp 0.64$.

We obtained the energy of neutrons by the time-of-flight (TOF) technique, and we used the kinematical conditions to select the $n + p$ events from the $n + \text{C}$ events: we compared the neutron velocity v_{exp} obtained from the TOF and the flight path length between counters with the velocity v_{NN} calculated from the incoming neutron energy assuming the $n + p$ kinematics. In order to obtain the left/right asymmetry, we applied the sector method[1] (Figure 2), where the neutrons with the scattering angles within the regions of $\theta_{\text{min}} < \theta < \theta_{\text{max}}$ and $|\phi| < \Phi$ ($|\phi - \pi| < \Phi$) were considered to be scattered in the right (left) direction.

The software cuts were optimized to obtain the maximum FOM: the events which satisfied $0.77 \leq v_{NN}/v_{\text{exp}} \leq 1.09$ and whose pulse height was $> 5.0\,\text{MeV}_{\text{ee}}$ were selected. The sector angles Φ and $(\theta_{\text{min}}, \theta_{\text{max}})$ were taken to be $66.8°$ and $(16.0°, 36.5°)$.

The results are $A_y^{\text{eff}} = 0.28 \pm 0.01$, $\epsilon_{\text{double}} = (4.16 \pm 0.01) \times 10^{-4}$, and FOM $= (3.2 \pm 0.1) \times 10^{-5}$, where the statistical uncertainties are denoted. The intrinsic energy resolution of NPOL is 1.0 MeV in FWHM.

References

1. M. Palarczyk et al.: Nucl. Instrum. Methods Phys. Res. A **457** 309 (2001).
2. H. Sakamoto et al.: Phys. Lett. B **155**, 227 (1985).

A FROZEN-SPIN TARGET FOR THE TOF DETECTOR

A. RACCANELLI*AND H. DUTZ

Institute of Physics, University of Bonn
Nussalle 12,
53115 Bonn, Germany

R. KRAUSE

Institute for Numerical Simulation, University of Bonn
Wegelerstr. 6,
53115 Bonn, Germany

The scattering experiments for which a high beam intensity and collimation are required demand challenging performance of a frozen-spin polarized target, both in terms of radiation resistance and of thermal transport properties. We are building a finite element model to study the temperature profile in target materials for the cases that do not allow an analytical solution or the use of approximation formulae.

For the experiment under preparation at COSY to measure the parity of the pentaquark $\Theta^+(1540)$, the requisites of luminosity and accuracy in the vertex reconstruction correspond to a beam intensity of 10^7 protons/s at a beam spot of about 1 mm². Under these conditions, ammonia and lithium compounds are the only materials that offer suitable radiation resistance[1]. Since localized beam heating causes inhomogeneous depolarization of the target and since the relaxation time quickly decreases with increasing temperature, the heat deposited by the beam must be efficiently dissipated toward the cryogenic bath. To reduce the effect of the Kapitza resistance on the boundary, the surface area of target must be maximized, e.g. by preparing the target as a series of thin slices of material. In this case, lithium compounds appear to be the most suitable choice, for the easier handling and preparation, since ammonia is in the gaseous state at room temperature. The choice of size and shape of the target, and in particular of its thickness, must be based on the evaluation of the temperature

---

*Corresponding author. E-mail: araccan@physik.uni-bonn.de

profile inside the target for the specific experimental conditions. Unfortunately, the thermal conductivity of irradiated lithium compounds at very low temperature is not known. Furthermore, if one uses the known thermal conductivity of butanol and the generally adopted approximations[2] to evaluate the temperature profile, the estimated temperature at the centre of the target turns out to be too high (fig. 1) and shows that the experiment is apparently unfeasibile[3].

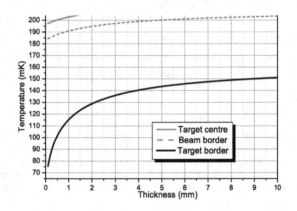

Figure 1. Temperature profiles as a function of the target thickness for a cylindrical target of 6 mm diameter at an operating temperature of 55 mK. For the lowest values of the thickness, the approximations used for the temperature estimate are no longer valid and lead to an overestimate of the temperature in the target.

Driven by the need for a more realistic way to estimate the temperature profile inside the target, we are building a new finite-element model to study the thermal transport properties as a function of size and shape of the target and of the thermal conductivity parameters. The implementation of the model is based on the UG package[4] and includes the Kapitza resistance as a boundary condition.

References
1. S. Goertz et al., *Prog. Part. Nuc. Phys.* **49**, 403-489 (2002).
2. M. Plückthun, *PhD-thesis*, Bonn (1998).
3. A. Raccanelli, in *Polarized Nucleon Targets for Europe*, Bochum (2004)
4. P. Bastian et al., *Comput. visual. sci.* **1**, 27-40 (1997).

POLARIZATION MEASUREMENT OF POLARIZED PROTON SOLID TARGET VIA THE $\vec{p}+^4$HE ELASTIC SCATTERING

S. SAKAGUCHI[a], T. UESAKA[a], T. WAKUI[b], T. KAWABATA[a], N. AOI[c],
Y. HASHIMOTO[d], M. ICHIKAWA[e], Y. ICHIKAWA[f], K. ITOH[g], M. ITOH[b],
H. IWASAKI[f], T. KAWAHARA[h], H. KUBOKI[f], Y. MAEDA[a], R. MATSUO[e],
T. NAKAO[f], H. OKAMURA[b], H. SAKAI[f], N. SAKAMOTO[c],
Y. SASAMOTO[a], M. SASANO[f], Y. SATOU[d], K. SEKIGUCHI[c],
M. SHINOHARA[d], K. SUDA[a], D. SUZUKI[f], Y. TAKAHASHI[f], A. TAMII[i],
K. YAKO[f], AND M. YAMAGUCHI[c]

[a] *Center for Nuclear Study, University of Tokyo, Bunkyo, Tokyo, Japan*
[b] *Cyclotron and Radioisotope Center, Miyagi, Japan*
[c] *RIKEN (The Institute of Physical and Chemical Research), Saitama, Japan*
[d] *Department of Physics, Tokyo Institute of Technology, Meguro, Tokyo, Japan*
[e] *Department of Physics, Tohoku University, Miyagi, Japan*
[f] *Department of Physics, University of Tokyo, Bunkyo, Tokyo, Japan*
[g] *Department of Physics, Saitama University, Saitama, Japan*
[h] *Department of Physics, Toho University, Chiba, Japan*
[i] *Research Center for Nuclear Physics, Osaka University, Osaka, Japan*

Recently, we have constructed a solid polarized proton target that can be used in RI beam experiments. The absolute value of the target polarization was measured via the $\vec{p}+^4$He elastic scattering at 80 MeV/nucleon. The average and the maximum polarization were 13.8 ± 3.9% and 20.4 ± 5.8%, respectively.

For the study of unstable nuclei, we have constructed a solid polarized proton target [1] which has a unique capability of operating in a low magnetic field of 80 mT and at high temperature of 100 K [2]. This modest operational condition enables us to use the target in RI beam experiments. However, the condition makes it difficult to determine the absolute value of the target polarization, because a proton polarization under thermal equilibrium which is required for the calibration of NMR signal is too small to detect under such a condition. Thus, as an alternative method, we measured the absolute polarization via the $\vec{p}+^4$He elastic scattering at 80 MeV/nucleon. Polarization can be deduced by dividing the measured asymmetry of the scattering by the analyzing power which has already been measured by Togawa et al. [3].

The experiment was carried out at RIKEN Accelerator Research Facility. The energy and the typical intensity of a ^{4}He beam were 80 MeV/nucleon and 250 kcps, respectively. The left panel of Fig. 1 shows the time evolution of the NMR signal amplitude, which is proportional to the proton polarization. Polarizing direction was reversed three times during the experiment for the compensation of the instrumental asymmetry.

The right panel of Fig. 1 shows a schematic view of the experimental setup. Scattering angles, energy deposits, and total energies of both protons and ^{4}He particles were measured by a multi wire drift chamber, an array of plastic scintillators, and a pair of counter telescopes placed left and right sides of the beam axis. Each telescope consists of a single wire drift chamber and a CsI (Tl) scintillator.

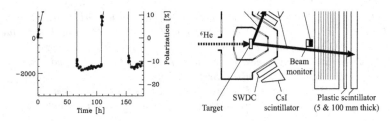

Figure 1. Time evolution of the polarization and the experimental setup are shown.

Using yields of the elastic scattering to the left (L) and right (R) directions in both cases where the polarizing directions are up ($\uparrow$) and down ($\downarrow$), the polarization P_y was deduced by $P_y = \frac{1}{A_y} \frac{\sqrt{N_L^\uparrow N_R^\downarrow} - \sqrt{N_R^\uparrow N_L^\downarrow}}{\sqrt{N_L^\uparrow N_R^\downarrow} + \sqrt{N_R^\uparrow N_L^\downarrow}}$, canceling the instrumental asymmetry. The NMR signal amplitude was then calibrated to give the absolute polarization as shown in the right-side vertical axis of the left panel of Fig. 1. The averaged polarization during the experiment was 13.8±3.9% and the maximum polarization was 20.4±5.8%.

References
1. T. Uesaka et al., Nucl. Instr. Meth. **A526**, 186 (2004).
2. T. Wakui et al., in this proceedings.
3. H. Togawa, RCNP Annual Report, 1 (1987).

A NEW TOOL TO CALIBRATE DEUTERON BEAM POLARIZATION AT INTERMEDIATE ENERGIES

K. SUDA[a], H. OKAMURA[b], T. UESAKA[a], J. NISHIKAWA[c],
H. KUMASAKA[c], R. SUZUKI[c], H. SAKAI[d], A. TAMII[e], T. OHNISHI[f],
K. SEKIGUCHI[f], K. YAKO[d], S. SAKODA[d], H. KATO[d], M. HATANO[d],
Y. MAEDA[a], T. SAITO[d], T. ISHIDA[d], N. SAKAMOTO[f], Y. SATOU[g],
K. HATANAKA[e], T. WAKASA[h] AND J. KAMIYA[e]

[a] *Center for Nuclear Study, University of Tokyo, Bunkyo, Japan*
[b] *Cyclotron and Radioisotope Center, Tohoku University, Sendai, Japan*
[c] *Department of Physics, Saitama University, Saitama, Japan*
[d] *Department of Physics, University of Tokyo, Bunkyo, Japan*
[e] *Research Center for Nuclear Physics, Osaka University, Ibaraki, Japan*
[f] *RIKEN (The Institute of Physical and Chemical Research), Wako, Japan*
[g] *Department of Physics, Tokyo Institute of Technology, Meguro, Japan*
[h] *Department of Physics, Kyushu University, Higashi, Japan*

The ^{12}C$(\vec{d},\alpha)^{10}$B*[2$^+$] reaction at 0° has been proposed as a new tool to calibrate the deuteron beam polarization at intermediate energies. The absolute values of the deuteron polarization are extracted for the first time by using this reaction at E_d =270, 200, and 140 MeV. The analyzing powers for the $\vec{d}$–p elastic scattering were determined to use as a secondary calibration reaction for a polarimetry.

1. Introduction

In recent years, extensive studies have been performed using polarized deuteron beams at intermediate energies (E_d >100 MeV). Particularly, the study of the three-nucleon forces (3NF) draws much attention [1]. In the study, highly precise data are required to be compared with rigorous calculations. For the reliable deduction of polarization observables, it is important to measure accurate beam polarization. At RIKEN, the $\vec{d}$–p elastic scattering is used to measure the beam polarization [2]. The analyzing powers for the $\vec{d}$–p elastic scattering were calibrated against the polarization measured by the low energy polarimeter [2, 3]. However, a systematic uncertainty arises from the fact that the beam polarization and the analyzing powers can not be measured simultaneously. Moreover, systematic uncertainties of the analyzing powers for the reaction used for the low energy polarimeter are already significant in amount. Thus, it is desired to calibrate the analyzing powers using the "calibration standard" reaction, the absolute value of which is unambiguously known. At low energy, a few calibration standard for deuteron beams is known, however, at intermediate energies, such a reaction has not been established.

As a practical calibration standard at intermediate energies, we propose to use the ^{12}C($\vec{d},\alpha$)^{10}B reaction. The tensor analyzing power A_{zz} at $\theta = 0°$ is identical to unity because of the parity conservation, if the residual nucleus ^{10}B* is in a natural parity state except for 0+. Among the levels of ^{10}B*, the 2+ state at E_x =3.59 MeV is advantageous because energy differences between the adjacent levels are larger than 1 MeV. In addition, even if the beam has a finite value of the vector polarization, it has no effect on the measurement of the tensor polarization because $A_y = 0$ at $\theta = 0°$.

2. Experiment

A calibration measurement was performed at the RIKEN Accelerator Research Facility using polarized deuteron beams at E_d=270, 200, and 140 MeV. The spin-dependent cross section for the ^{12}C($\vec{d},\alpha$)^{10}B*[2+] reaction was measured at $\theta = 0°$ using the spectrograph SMART [4]. Simultaneously, using the same beam, the asymmetries of the $\vec{d}$–p elastic scattering were measured at six different angles within $\theta_{c.m.} = 80°$–$120°$ using a beam line polarimeter.

3. Results

The absolute values of the beam polarization were deduced by using the ^{12}C($\vec{d},\alpha$)^{10}B*[2+] reaction. Using the obtained beam polarization, the analyzing powers for the $\vec{d}$–p elastic scattering were calibrated with high accuracy. The combined statistical and systematic errors are within ±0.02 for A_y, A_{yy}, and A_{xx}, and ±0.06 for A_{xz}, respectively. The present result is almost consistent with the previous calibrations [5].

References

1. H. Sakai et al., Phys. Rev. Lett. **84**, 5288 (2000).
2. N. Sakamoto et al., Phys. Lett. B **367**, 60 (1996).
3. T. Uesaka et al., RIKEN Accel. Prog. Rep. **33**, 153 (2000).
4. T. Ichihara et al., Nucl. Phys. **A569**, 287c (1994).
5. K. Sekiguchi et al., Phys. Rev. C **70**, 014001 (2004).

POLARIZED ³HE ION SOURCE BASED ON THE SPIN-EXCHANGE COLLISIONS

M. TANAKA*

*Dept. of Clinical Technology, Kobe Tokiwa College
Ohtani-cho 2-6-2, Nagata-ku, Kobe 653-0838, JAPAN
E-mail: tanaka@rcnp.osaka-u.ac.jp*

Y. TAKAHASHI, T. SHIMODA, AND T. FURUKAWA

*Dept. of Physics, Graduate School, Osaka University,
Machikaneyama-cho1-1, Toyonaka, Osaka 560-0043, JAPAN*

S. YASUI,[†] M. YOSOI, AND K. TAKAHISA

*Research Center for Nuclear Physics, Osaka University,
Mihogaoka 10-1, Ibaraki, Osaka 567-0047, Japan*

A new type of polarized ³He ion source is developed in collaboration with RCNP and Dept. of Physics, Osaka University. The principle to produce the polarized beams is based on an extended concept of the OPPIS (Optical Pumping Polarized Ion Source) succeeded in efficiently polarizing proton beams, *i.e.*, an enhanced spin-exchange cross section between a ³He⁺ ion and a polarized alkali atom (Rb) at incident energies lower than a few keV.

1. Introduction

Polarized ³He beams at intermediate and high energy regions provide one of the important tools not only in nuclear physics but also in particle physics. Recently, our group proposed a novel idea to achieve production of the polarized ³He²⁺ beams with high intensity and high polarization [1]. The principle of polarization is to use an unexpectedly large spin-exchange cross section($\geq 10^{-14}$ cm²) and a small electron capture cross section between

---

*Work partially supported by the Grant-in-Aid (no.16540720) by the Japan Ministry of Education, Culture, Sports, Science, and Technology.
†Present Address: Dept. of Physics, Tokyo Institute of Technology, O-okayama 2-12-1, Tokyo 152-8551, Japan

a ^{3}He$^+$ ion and a Rb atom at low incident energy regions ($\leq$ a few keV). We name the polarized ion source based on this principle "SEPIS", i.e., Spin-Exchange Polarized Ion Source. The aim of the present work is to experimentally prove validity of the SEPIS and to examine feasibility toward a practical polarized ^{3}He ion source with high polarization and high intensity. For this purpose, we currently prepare the measurements of the spin-exchange cross sections and electron capture cross sections for a ^{3}He$^+$ and a Rb atom in a broad energy region down hopefully to sub-keV.

2. Equipment

The construction of a bench-test device shown in Fig.1 is almost in completion. An unpolarized ^{3}He$^+$ ion is produced by a 2.45GHz ECR ion source with a mirror field and multipole field generated by permanent magnets. The 19 keV ^{3}He$^+$ ion extracted from the ion source is introduced to an Rb vapor cell after momentum analysis with a bending magnet. To enable the spin-exchange collisions at low incident energies ($\leq$ a few keV) a high voltage (up to $\sim$ 19 kV) is applied to the Rb cell electrically insulated. The polarized ^{3}He$^+$ ion emerging out of the Rb cell, where the multiple spin-exchange collisions occur between the ^{3}He$^+$ ion and Rb atoms polarized by the optical pumping, is energy-analysed by an electrostatic analyzer with spherical electrodes after the ^{3}He$^+$ ion is accelerated up to the primary 19 keV again. The polarized ^{3}He$^+$ ion is, then, introduced to a polarimeter

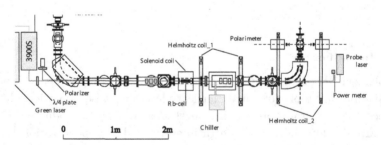

Figure 1. A schematic layout of the bench-test device for proving validity of the SEPIS principle.

where the generated nuclear polarization is measured by means of the beam foil spectroscopy. A holding magnetic field ($\sim$ 10 G) produced by two sets of Helmholtz coils covers a whole region of the equipment downstream from the exit of the Rb vapor cell to the device end so that the depolarization due to perturbing magnetic fields like a terrestrial magnetism might be avoided.

3. Experimental results and discussion

3.1. ECR ion source

To optimize the performance of the ECR ion source we have measured ^{3}He$^+$ ion currents by varying ^{3}He gas flow rate, extraction voltage, 2.45 GHz microwave power. The ^{3}He$^+$ beam current over 200 eμA is easily achieved, whose value is enough large to precisely measure the spin exchange cross sections.

3.2. Electron capture cross section

We measured the electron capture cross sections in a ^{3}He$^+$ energy range from 19 down to 1 keV to investigate the expected ^{3}He$^+$ beam current. The capture cross sections were obtained by taking a ratio of ^{3}He$^+$ ion current emerging out of the Rb cell to that incident on the Rb cell as a function of Rb vapor thickness which was measured by a Faraday rotation angle of a 782-nm probe laser penetrating the Rb vapor in the presence of magnetic field.

3.3. Polarimeter based on the beam foil spectroscopy

Performance of the polarimeter was achieved by using atomically polarized ^{3}He I created by the tilting foil method. An important result is that the data are reproducible enough precise to test the validity of SEPIS requiring the measurements of ^{3}He nuclear polarization better than 0.5 %.

4. Future prospect

Installation and tuning of the pumping laser system consisting of a 10W green laser and a Ti:Sapphire laser was almost completed and the measurement of spin-exchange cross sections will soon get started. After proving the SEPIS principle, we will start designing and constructing a practical polarized ^{3}He^{2+} ion source.

References

1. M. Tanaka *et al.* Nucl. Instr. Meth. **A537** (2005) 501-509.

EXTRACTION OF FRACTIONS OF THE RESONANT COMPONENT FROM ANALYZING POWERS IN ^{6}LI(D, α)^{4}HE AND ^{6}LI(D, P$_0$)^{7}LI REACTION AT VERY LOW INCIDENT ENERGIES

M. YAMAGUCHI AND Y. TAGISHI

RIKEN, Wako, Saitama, Japan

Y. AOKI, T. IIZUKA, T. NAGATOMO, T. SHINBA,
N. YOSHIMARU AND Y. YAMATO

Institute of Physics and Tandem Accelerator Center, University of Tsukuba, Ibaraki, Japan

T. KATABUCHI

Gunma University Graduate School of Medicine, Gunma, Japan

M. TANIFUJI

Department of Physics, Hosei University, Tokyo, Japan

We measured the analyzing powers of polarized deuterons at the incident energy of 90 keV in the ^{6}Li(d, p$_0$)^{7}Li (g.s.) and ^{6}Li(d, α)^{4}He (g.s.) reactions and extracted the fraction of the resonant component (resonance state of ^{8}Be, $E_x = 22.2$MeV, $\Gamma \sim 800$keV, $J^\pi = 2^+$). The experiment was performed using a Lamb shift-type polarization ion source [1] at the University of Tsukuba Tandem Accelerator Center (UTTAC). Polarized deuteron beam was introduced to a lithium target. Twelve Si-SSD's are placed around the target at every 15° scattering angles from 0° to 165° to detect the emitted proton and α-particles. Figure 1 shows the measured analyzing powers for (d, p$_0$) and (d, α) reactions. The predictions from IMA (Invariant Amplitude Method) [2] well reproduce the measured analyzing powers. From these data, we confirmed the spin and parity of the resonance to be 2^+ and obtained the fraction of the resonant component for the (d, p$_0$) reaction as 0.90 ± 0.05 which is consistent with the earlier value about 0.85 extracted from the cross section data. As the fraction of the resonant component for the (d, α) reaction, we obtained 0.998 ± 0.003, which is much larger than the earlier value about 0.5 extracted from the cross section data.

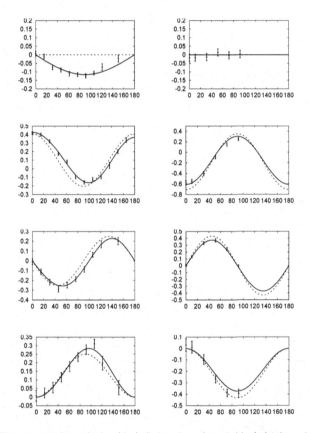

Figure 1. Experimental result for (d,p) (left column) and (d,α) (right column). The lines are the prediction of IMA. For (d,p), the dashed lines are the results in which the angular momentum of the incident channel is restricted to s-wave. For (d,α), the dashed lines are the results in which total angular momentum are restricted to 2^+. The abscissas represent the scattering angle in the C.M. system.

References

1. Y. Tagishi and J. Sanada *Nucl. Instr. and Meth.* **164**, 411 (1979).
2. M. Tanifuji and H. Kameyama *Phys. Rev.* **C60**, 034607 (1999).

Participants List

Takamasa Arai
Tokyo Institute of Technology
2-12-1, Ookayama, Megro-ku, Tokyo, 152-8550, Japan
arai@yap.nucl.ap.titech.ac.jp

Koichiro Asahi
RIKEN/TITech
2-1 Hirosawa, Wako, Saitama 351-0198, Japan
asahi@phys.titech.ac.jp

Anatoly Beda
Institute for Theoretical and Experimental Physics
b. Cheremushkinskaya 25, Moscow 117218, Russia
beda@itep.ru

Sylvain Bouchigny
IPN Orsay
F-91406 ORSAY CEDEX, France
bouchign@ipno.in2p3.fr

Axel Brachmann
Stanford Linear Accelerator Center
2575 Sand Hill Road, Menlo Park, CA 94025, USA
brachman@slac.stanford.edu

Ben Clasie
Mass. Inst. of Tech.
26-650b 77 Massachusetts Ave Cambridge MA 02139 USA
clasie@mit.edu

James E. Clendenin
Stanford Linear Accelerator Center
2575 Sand Hill Road, Menlo Park, CA 94025, USA
clen@slac.stanford.edu

D. G. Crabb
University of Virginia
Physics Dept., 382 McCormick Rd., Charlottesville, VA 22903, USA
dgc3q@virginia.edu

Chaden Djalali
University of South Carolina - Department of Physics and Astronomy
712 Main Street, Columbia, SC 29208, USA
djalali@sc.edu

Jean-Pierre, Didelez
Institut de Physique Nucleaire, Bat100
F-91406 ORSAY CEDEX, France
didelez@ipno.in2p3.fr

Norihiro Doshita
University of Bochum
PH Dep. CERN, CH-1211, Geneva 23, Switzerland
norihiro.doshita@cern.ch

Ralf Engels
Institut for Nuclear Physics, FZ Juelich, Germany
Leo-Brandt-Str. 1, 52425 Juelich
r.w.engels@fz-juelich.de

Manouchehr Farkhondeh
MIT-Bates
26 Manning Road, Middleton, MA 01949, USA
manouch@mit.edu

Victor Fimushkin
Joint Institute for Nuclear Research
141980 Dubna, Moscow region, Russia
fimush@sunhe.jinr.ru

Michael Finger
CTU Prague
Sluknovska 312, CZ-19000 Praha, Czech Republic
michael.finger@cern.ch

Miroslav Finger
Charles University in Prague
V Holesovickach 2, CZ-18000 Praha, Czech Republic
miroslav.finger@cern.ch

Takeshi Furukawa
Department of Physics, Graduate School of Science, Osaka University
1-1, Machikaneyama, Toyonaka, Osaka, 560-0043, Japan
take@valk.phys.sci.osaka-u.ac.jp

Gvirol Goldring
The Weizmann Institute of Science
Rehovot 76100 Israel
gvirol.goldring@weizmann.ac.il

Katerina Ioakeimidi
Stanford Linear Accelerator Center
2575 Sand Hill Road, Menlo Park, CA 94025, USA
katerina@slac.stanford.edu

Bertalan Juhasz
Stefan Meyer Institute for Subatomic Physics of the Austrian Academy of Sciences
Boltzmanngasse 3, 1090 Wien, Austria
bertalan.juhasz@assoc.oeaw.ac.at

Kichiji Hatanaka
Research Center for Nuclear Physics, Osaka University
10-1 Mihogaoka, Ibaraki, Osaka 567-0047, Japan
hatanaka@rcnp.osaka-u.ac.jp

Patrick Hautle
Paul Scherrer Institute
5232 Villigen PSI, Switzerland
patrick.hautle@psi.ch

Naoaki Horikawa
College of Engineering, Chubu University
Kasugai, Aichi 487-8501, Japan
horikawa@isc.chubu.ac.jp

Keisuke Itoh
Saitama University
255, Shimo-ohkubo, Sakura-ku,Saitama,Japan
keisuke@ne.phy.saitama-u.ac.jp

Takahiro Iwata
Yamagata University
Kojirakawa-chou 1-4-12, Yamagata, 990-8560, Japan
tiwata@sci.kj.yamagata-u.ac.jp

Tsuneo Kageya
Brookhaven National Lab. and Virginia Tec
LEGS group, Bld. 510A, Brookhaven National Lab., Upton, NY 11973-5000, USA
kageya@bnl.gov

Daisuke Kameda
RIKEN
2-1, Hirosawa, Wako, Saitama, 351-0198, Japan
kameda@rarfaxp.riken.jp

Tatsuya Katabuchi
Gunma University
3-39-22 Showa, Maebashi, Gunma 371-8511, Japan
buchi@taka.jaeri.go.jp

Takahiro Kawabata
Center for Nuclear Study, University of Tokyo
7-3-1 Hongo, Bunkyo, Tokyo, 113-0033, Japan
kawabata@cns.s.u-tokyo.ac.jp

Tomomi Kawahara
Toho University
158-4 Kamiterayama, Kawagoe, Saitama 350-0826
kawahara@cns.s.u-tokyo.ac.jp

Go Kijima
Department of Physics, Tokyo Institute of Technology (Asahi Lab.)
2-12-1 Ookayama, Meguro-ku, Tokyo, 152-8550, JAPAN (Office N.O. 412, South 5th building, T.I.T)
kijima@yap.nucl.ap.titech.ac.jp

Yuri Kisselev
CERN
CERN/PH, Bat.892 1-D12, 1211 Geneva 23 Switzerland
Yuri.Kisselev@cern.ch

Jaakko Koivuniemi
CERN/PH
CH-1211 Genève 23, Switzerland
Jaakko.Koivuniemi@cern.ch

Jochen Krimmer
University Mainz
Staudinger-Weg 7, Germany
krimmer@uni-mainz.de

Hironori Kuboki
Department of Physics, University of Tokyo
Hongo 7-3-1, Bunkyo, Tokyo, Japan
kuboki@nucl.phys.s.u-tokyo.ac.jp

Takayuki Kumada
Japan Atomic Energy Agency
Tokai, Ibaraki 319-1195, Japan
kumada.takayuki@jaea.go.jp

Kazuyoshi Kurita
Rikkyo University
3-34-1 Nishi-Ikebukuro Tokyo 171-8501, Japan
kurita@ne.rikkyo.ac.jp

Makoto Kuwahara
Nagoya University
464-8602, Furo-cho Chikusa-ku Nagoya-city Aichi-prefecture Japan
kuwahara@spin.phys.nagoya-u.ac.jp

Ladygin Vladimir
Laboratory of High Energies of Joint Institute for Nuclear Research
141980, Joliot Curie 6, LHE-JINR, Dubna, Moscow region, Russia
ladygin@sunhe.jinr.ru

Yukie Maeda
Center for Nuclear Study, University of Tokyo
2-1 Hirosawa, Wako, Saitama 351-0198, Japan
yukie@nucl.phys.s.u-tokyo.ac.jp

Takashi Maruyama
SLAC
2575 Sand Hill Road, M/S 74
tvm@slac.stanford.edu

Akira Masaike
Washington Center, Japan Society for the Promotion of Science
1800 K Street, NW, Suite 920, Washington DC, 20006, USA
masaike@jspsusa.org

Yasuhiro Masuda
Institute of Particle and Nuclear Studies, KEK
1-1 Oho, Tsukuba-shi, Ibaraki 305-0801, Japan
yasuhiro.masuda@kek.jp

Werner Meyer
Ruhr-Universität Bochum
Universitätsstr. 150, Bochum 44780, Germany
meyer@ep1.rub.de

Mototsugu Mihara
Department of Physics, Graduate School of Scinence, Osaka University
1-1, Machikaneyama, Toyonaka, Osaka 560-0043, Japan
mihara@vg.phys.sci.osaka-u.ac.jp

Kenjiro Miki
Department of Physics, University of Tokyo
7-3-1 Hongo, Bunkyo, Tokyo, 113-0033
miki@nucl.phys.s.u-tokyo.ac.jp

Hiroari Miyatake
Institute of Particle and Nuclear Studies, KEK
1-1 Oho, Tsukuba-shi, Ibaraki 305-0801, Japan
hiroari.miyatake@kek.jp

Takashi Nakajima
Institute of Advanced Energy, Kyoto University
Gokasho, Uji, Kyoto 611-0011, Japan
t-nakajima@iae.kyoto-u.ac.jp

Tsutomu Nakanishi
Department of Physics, Graduate School of Science, Nagoya University
Furo-cho, Chikusa-ku, Nagoya-city, 464-8602
nakanisi@spin.phys.nagoya-u.ac.jp

Itaru Nishikawa
Laboratory of Nuclear Science, Tohoku University
1-2-1, Mikamine, Taihaku-ku, Sendai city, Japan
nisikawa@lns.tohoku.ac.jp

Shumpei Noji
Department of Physics, University of Tokyo
7-3-1 Hongo, Bunkyo, Tokyo, 113-0033
noji@nucl.phys.s.u-tokyo.ac.jp

Hiroyuki Okamura
Cyclotron & Radioisotope Center (CYRIC), Tohoku University
6-3 Aoba, Aramaki, Aoba, Sendai 980-8578, Japan
okamura@cyric.tohoku.ac.jp

Matt Poelker
Jefferson Lab
12000 Jefferson Avenue, MS 5A, USA
poelker@jlab.org

Andrea Raccanelli
Physikalisches Institut, Universität Bonn
Nussallee 12, D-53115, Germany
araccan@physik.uni-bonn.de

Michael Romalis
Department of Physics, Princeton University
Princeton, NJ 08544, USA
romalis@princeton.edu

Naohito Saito
Department of Physics, Kyoto University
Sakyo-ku, Kyoto 606-8502, Japan
saito@nh.scphys.kyoto-u.ac.jp

Satoshi Sakaguchi
Center for Nuclear Study, University of Tokyo
2-1 Hirosawa, Wako, Saitama, 351-0198 Japan
satoshi@cns.s.u-tokyo.ac.jp

Hideyuki Sakai
Department of Physics, University of Tokyo
7-3-1 Hongo, Bunkyo, Tokyo, 113-0033, Japan
sakai@phys.s.u-tokyo.ac.jp

Yoshiko Sasamoto
Center for Nuclear Study, University of Tokyo
2-1 Hirosawa, Wako, Saitama 351-0198, Japan
yoshiko@cns.s.u-tokyo.ac.jp

Masaki Sasano
Department of Physics, University of Tokyo
7-3-1 Hongo, Bunkyo-ku, Tokyo
sasano@nucl.phys.s.u-tokyo.ac.jp

Kimiko Sekiguchi
RIKEN
Hirosawa 2-1, Wako, Saitama 351-0198, JAPAN
kimiko@rarfaxp.riken.jp

Youhei Shimizu
Research Center for Nuclear Physics, Osaka University
10-1 Mihogaoka, Ibaraki, Osaka 567-0047, Japan
yshimizu@rcnp.osaka-u.ac.jp

Tadashi Shimoda
Department of Physics, Graduate School of Science, Osaka University
1-1 Machikaneyama, Toyonaka, Osaka 560-0043, Japan
shimoda@phys.sci.osaka-u.ac.jp

Edward J. Stephenson
Indiana University Cyclotron Facility
2401 Milo B. Sampson Lane, Bloomington, IN 47408, USA
stephens@iucf.indiana.edu

Kenji Suda
Center for Nuclear Study, University of Tokyo
7-3-1 Hongo, Bunkyo-ku, Tokyo, 113-0033 Japan
suda@cns.s.u-tokyo.ac.jp

Hideo Suzuki
Spectra Gas Inc.
6-1-407 Shimo-Namiki, Kawasaki-ku, Kawasaki, Japan
suzukitrd@yahoo.co.jp

Yoshiyuki Takahashi
University of Tokyo
7-3-1, Hongo,Bunkyo-ku, Tokyo 113-0033, Japan
ytakahashi@nucl.phys.s.u-tokyo.ac.jp

Makoto Takemura
Tokyo Institute of Technology
2-12-1 Ookayama, Meguro-ku,
takemura@yap.nucl.ap.titech.ac.jp

Tadaaki Tamae
Tohoku University
Mikamine, Taihaku-Ku Sendai
tamae@lns.tohoku.ac.jp

Atsushi Tamii
Research Center for Nuclear Physics, Osaka University
10-1, Mihogaoka, Ibaraki 567-0047, Japan
tamii@rcnp.osaka-u.ac.jp

Masayoshi Tanaka
Kobe Tokiwa College
Ohtani-cho 2-6-2, Nagata-ku, Kobe 653-0838, Japan
tanaka@rcnp.osaka-u.ac.jp

Evgeni P. Tsentalovich
MIT
21 Manning Rd, Middleton, MA, USA, 01949-2846
evgeni@mit.edu

Makoto Uchida
Department of Physics, Tokyo Institute of Technology
Meguro-ku, Tokyo 152-8551, Japan
uchida@phys.titech.ac.jp

Hideki Ueno
RIKEN
2-1 Hirosawa, Wako, Saitama 351-0198, Japan
ueno@riken.jp

Tomohiro Uesaka
Center for Nuclear Study, University of Tokyo
7-3-1 Hongo, Bunkyo, Tokyo, 113-0033, Japan
uesaka@cns.s.u-tokyo.ac.jp

Takashi Wakui
Cyclotron & Radioisotope Center (CYRIC), Tohoku University
6-3 Aoba, Aramaki, Aoba, Sendai 980-8578, Japan
wakui@cyric.tohoku.ac.jp

Thomas Wise
Department of Physics, University of Wisconsin
1150 university Avenue, Madison, WI 53706, USA
wise@physics.wisc.edu

Kentaro Yako
Department of Physics, University of Tokyo
7-3-1 Hongo, Bunkyo, Tokyo, 113-0033, Japan
yakou@phys.s.u-tokyo.ac.jp

Mitsutaka Yamaguchi
RIKEN (Heavy Ion Nuclear Physics Lab.)
Hirosawa 2-1, Wako-shi, Saitama, Japan
myamagu@rarfaxp.riken.jp

Masahiro Yamamoto
Nagoya University
Furo-cho, Chikusa-ku, Nagoya, Japan
yamamoto@spin.phys.nagoya-u.ac.jp

Akihiro Yoshimi
RIKEN
2-1 Hirosawa, Wako, Saitama 351-0198, Japan
yoshimi@rarfaxp.riken.go.jp

Anatoli Zelenski
BNL
Bldg.930, Brookhaven National Laboratory
zelenski@bnl.gov